IN THE WAKE OF BERNARD HEUVELMANS

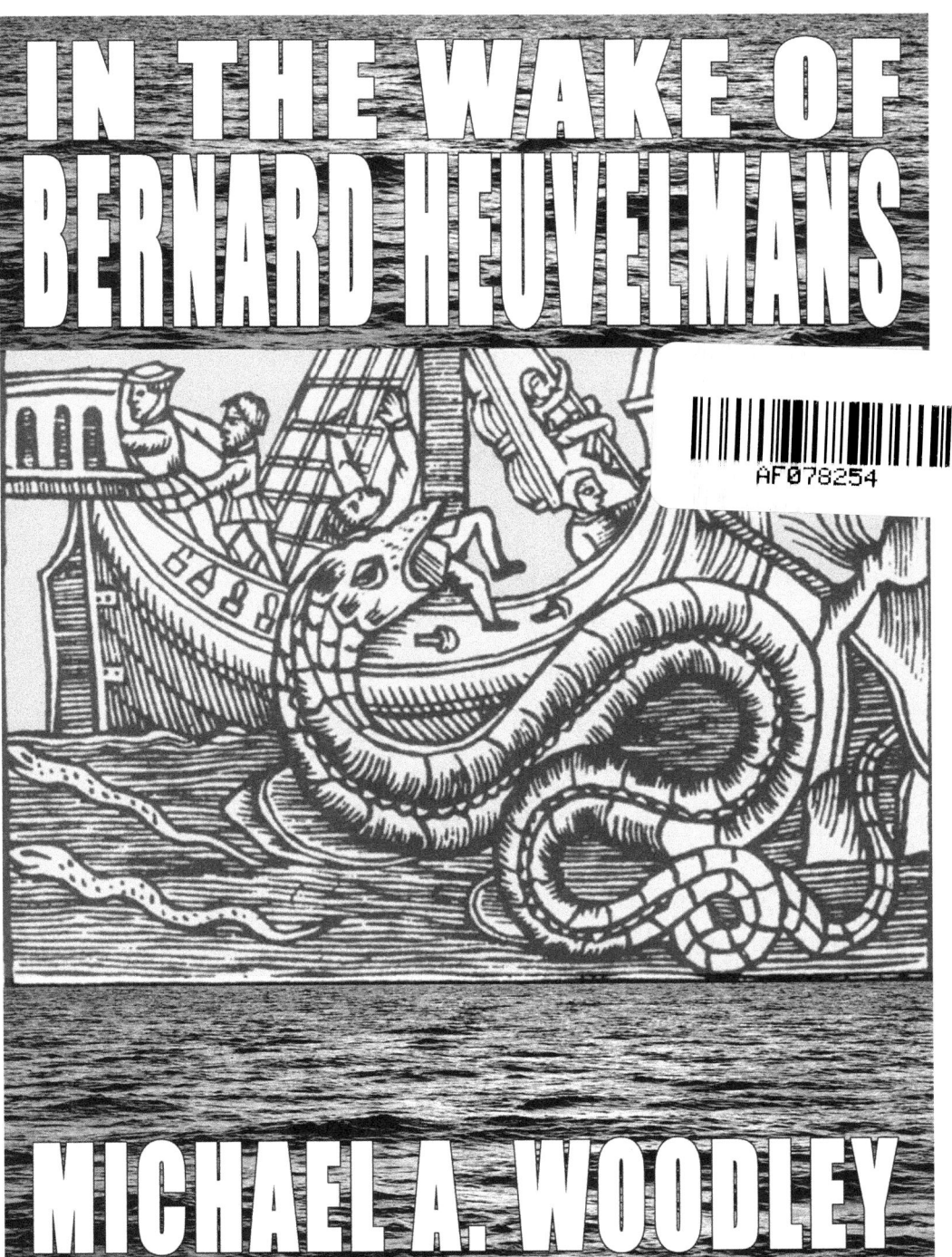

MICHAEL A. WOODLEY

Typeset by Jonathan Downes,
Cover and Layout by Hennis and *Felis oranjiensis* for CFZ Communications
Using Microsoft Word 2000, Microsoft , Publisher 2000, Adobe Photoshop CS.

Photographs © 2008 CFZ except where noted

First published in Great Britain by CFZ Press

**CFZ Press
Myrtle Cottage
Woolsery
Bideford
North Devon
EX39 5QR**

© CFZ MMVIII

All rights reserved. Without limiting the rights under copyright reserved above, no part of this publication may be reproduced, stored in or introduced into a retrieval system, or transmitted, in any form of by any means (electronic, mechanical, photocopying, recording or otherwise), without the prior written permission of both the copyright owners and the publishers of this book.

ISBN: 978-1-905723-20-1

For the CFZ - keep up the good work

Contents

11. Foreword By Dr. Karl P.N. Shuker
13. **An introduction to the history and future of sea serpent classification.**
    1. INTRODUCTION
       1.1. A long history
14. 1.2 The first rigorous sea serpent classification scheme: The Rafinesque model
17. 1.3 Nineteenth century enthusiasm: Oudemans' Megophias megophias
18. 1.4 Everything changes: The Heuvelmans model
21. 1.5 Post-Heuvelmans: The Coleman & Huyghe model
    1.6 The future of aquatic cryptozoology: The Champagne model
23. 2. COMPARATIVE METHODOLOGIES
    2.1 Overview
    2.2 The integrative methodology
    2.2.1 Ethnozoological and Zoomythological evidence
25. 2.3 The multifactor methodology
    2.4 The plausibility methodology
27. 3. CONCLUSIONS
    3.1 Comparisons
28. 3.2 And finally, the object of the manuscript
31. **CHAPTER 1. Heuvelmans' Pinnipeds**

31.	1. INTRODUCTION
31.	1.1 Overview
32.	1.2 Identity theories
	1.2.1 The 'Elasmosaurus' theory
33.	1.2.2 The Basilosaurus theory
	1.2.3 The Giant squid theory
34.	1.2.4 The long necked seal theory
35.	1.3 Which is the more likely explanation?
39.	2. AN EVOLUTIONARY HISTORY OF THE PINNIPEDS
	2.1 Overview
	2.2 Taxonomy
	2.3 Polyphyletic or Monophyletic?
41.	2.4 Ecology
	2.5 Reproduction
	2.6 Key differences
43.	3. THE NATURE OF THE BEAST
	3.1 Formally introducing the longnecks
45.	3.2 Convergent evolution
	3.2.1 The niche of the elasmosaurus
46.	3.2.2 How longnecks and elasmosaurs seem to differ
49.	3.2.3 Long necks in other pinnipeds
	3.3 Migration patterns and habitats
50.	3.3.1 Competition between longneck species
	3.3.2 Competition with other species
53.	3.4 Longnecks as *another* pinniped family?
54.	3.5 Cryptic lifestyles
	3.6 Conclusions
55.	4. SIGHTINGS
	4.1 Overview
	4.2 Close encounters
	4.2.1 Encounter number one
57.	4.2.2 Encounter number two
58.	4.2.3 Encounter number three
60.	4.2.4 Encounter number four
62.	4.2.5 Encounter number five
63.	4.2.6 Encounter number six
65.	4.2.7 Encounter number seven
66.	4.2.8 Encounter number eight
68.	4.3 Summary
71.	5. WHAT DOES THE FUTURE HOLD IN STORE?
	5.1 Overview
	5.2 Conservation issues
72.	5.2.1 Over-fishing
	5.2.2 Increasing ocean traffic
	5.2.3 Pollution
73.	5.2.4 Climate change

73.   5.3 A conservation plan
      5.4 Chances of discovery
75. **CHAPTER 2. Heuvelmans' archaeocetes**
      The many finned sea serpent: Time to reclassify?
      1. INTRODUCTION
      1.1 Overview
76.   1.2 A centipede by any other name
77.   2. IDENTITY THEORIES.
      2.1 Overview
      2.1.1 The many-finned cetacean theories
78.   2.1.2 Invertebrate theories
81.   3. SIGHTINGS AND CARCASSES
      3.1 Descriptions of physical specimens
      3.1.1 Encounter number one
82.   3.1.2 Encounter number two
85.   4. SUGGESTING AN IDENTITY
      4.1 The case against a cetacean identity
      4.1.1 A polychaete identity?
86.   4.1.2 Or, a Myriapoda identity?
      4.2 Evolving the *con rit*
87.   4.3 Does it fit with the facts?
      4.4 A need for reclassification
89. **The super-otter and the many-humped sea serpent: close cousins?**
      1. INTRODUCTION
      1.1 Overview
      1.2 Descriptions
      1.3 The super-otter
90. **1.4 The many-humped sea serpent**
93.   2. IDENTITY THEORIES
      2.1 Heuvelmans' proposed identities
      2.1.1 Is it consistent?
      2.1.2 Archaeocetes
95.   3. SUPER-OTTERS AND MANY HUMPED SEA SERPENTS AS LUTRINAE
      3.1 Overview
      3.2 More alike than different
      3.2.1 A lutrinae identity for the super-otter
96.   3.2.2 A lutrinae identity for the many-humped sea serpent
97.   3.3 Their standing within the sub-family lutrinae
99.   4. SIGHTINGS
      4.1 Overview
      4.2 Encounter number one
100. 4.3 Encounter number two
103. 5. CONCLUSIONS
      5.1 A third lutrinae cryptid?
      5.2 Summary

- 105. **CHAPTER 3. The Others**
  Marine 'saurians': genuine archaeocetes?
  1. INTRODUCTION
  1.1 Overview
- 106. 1.2 Description
- 107. 2. IDENTITY THEORIES
  2.1 Thalattosuchians and mosasaurs
  2.1.1 Thalattosuchia
  2.1.2 Mosasaurs
- 108. 2.2 Crocodile-like archaeocetes
  2.3 Comparisons
- 111. 3. SIGHTINGS
  3.1 Overview
  3.2 Encounter number one
- 112. 3.3 Encounter number two
- 113. 3.4 Encounter number three
- 115. 4. CONCLUSIONS
- 117. **Super-eels: a many faceted enigma**
  1. INTRODUCTION
  1.1 Overview
  1.2 Description
- 119. 2. IDENTITY THEORIES
  2.1 Overview
  2.1.1 Frilled sharks
- 120. 2.1.2 Oarfish
  2.2 Comparisons
- 123. 3. A CURIOUS SIGHTING
  3.1 Overview
  3.2 The sighting
- 124. 3.3 Considerations
- 125. 4. CONCLUSIONS
- 127. **Giant invertebrates: The most plausible category.**
  1. INTRODUCTION
  1.1 Overview
  1.2 Description
- 129. 2. IDENTITY THEORIES
  2.1 Disentangling the giant invertebrates
  2.1.1 Urochordata – salp
  2.1.2 Ctenophora – Venus' girdle
- 130. 2.1.3 Cnidaria – The Arctic lion's mane jellyfish
  2.1.4 Nemertea – The bootlace worm
  2.2 Discussion
- 133. 3. CONCLUSIONS
- 135. End notes.
  1. SUMMARY

135. 1.1 Overview
136. 1.2 Comparative table of identity
137. 1.3 'Dropped' categories
     1.4 Additional categories and future research directions
     1.5 Fossil evidence and ghost lineages.
138. 2. AFTERWORD
139. Appendix: Provided here are more detailed descriptions of the various extinct taxa mentioned in the text.
151. Sources and further reading

Foreword

By Dr. Karl P.N. Shuker

This is not a book for those who naively assume - or narrow-mindedly assert - that cryptozoology is one of those lightweight, bubblegum "Gee whiz, look what's out there" subjects, built upon a shaky foundation of fanciful assumptions and little else. It certainly isn't the bedtime, dip-into crypto-coffee style of book that pervades the current publishing ethos like a choking, tenacious strain of waterweed, packed with glossy photos and minimalist text.

On the contrary, what we have here is an emphatically serious, sober examination of the history and philosophy of sea serpent identification and classification, with especial emphasis upon Dr Bernard Heuvelmans's system, in which the author presents his own, radically new version, offering some eye-opening, novel identities for various sea serpent types.

It is common knowledge that my own view of cryptozoology is that any attempt to identify cryptids, though an intriguing and undeniably entertaining exercise, should be recognised as being of secondary importance to seeking the beasts themselves in the field. No amount of armchair theorizing and pontificating can substitute for on-site investigation - something that the CFZ expeditions have proven time and again in recent years. Far too much emotional, even vitriolic condemnation of fellow researchers' suggestions as to what a given cryptid might - or might not - be has appeared in print and online, when in reality no theory, however, rational or seemingly plausible, can ever become more than a theory until physical evidence is obtained for direct analysis.

Nevertheless, there is definitely room for reasoned, objective consideration in relation to putative cryptid identities, and this book offers that in admirable fashion. As an intellectual exercise, assessing the possible taxonomic status of a given cryptid can be very rewarding, and may also unveil hitherto-unsuspected insights into the nature of the beast, making such analysis especially worthwhile. Consequently, I recommend that everyone with an interest in mystery beasts, particularly those of a marine persuasion, should read and consider the intriguing thoughts and original theories put forward in this new cryptozoological volume, the first of what I feel sure will be a noteworthy output of work from a fresh voice in this exciting scientific field.

Karl Shuker
West Midlands
29 February 2008.

EDITOR'S NOTE: Footnotes marked with a † have been taken from Wikipedia, the free online encyclopaedia. Although we are aware of the controversies which surround Wikipedia, we are loath to join the ranks of those who constantly belittle it. We believe that Wikipedia is a brave - if sometimes flawed - project, and would like to openly support it. These footnotes have been inserted purely on the decision of the editor. All numbered footnotes are the work of the author. JD

An introduction to the history and future of sea serpent classification.

1. INTRODUCTION

1.1. A long history

Everyone is familiar with the nautical maps from the middle ages that were liberally festooned with images of exotic and monstrous animals, but the truth of the matter is that the *idea* of the sea monster is probably as old as humankind itself.

When humanity's earliest ancestors looked out upon the oceans, what lurked beneath the seemingly impenetrable blue surface could only have been guessed at. Naturally as humans became bolder and took to the seas in primitive vessels, whales, sharks, and other hazards would have been encountered, which would have served to stimulate the imagination and generate legends.

As the scope of humanity's exploration grew, so too did the scope of its understanding of what the oceans contained, and indeed as the modern era dawned, naturalists became increasingly confident in the idea that they could account for every large animal in existence - an idea that found its ultimate expression in the form of the French naturalist Georges Cuvier's so called 'rash dictum' of 1821[†]. There were those who disagreed however.

1.2 The first rigorous sea serpent classification scheme: The Rafinesque model

In the November 1819 edition of the *Philosophical Magazine and Journal*, the brilliant but neurotic Naturalist Constantine Samuel Rafinesque-Schmaltz, published his *Dissertation on Water-Snakes, Sea Snakes, and Sea Serpents* in which he attempted to classify sea serpent sightings into four identity categories based loosely around the idea that they were new species of sea snake or eel. Each category was somewhat ambitiously granted its own scientific name. [††]

Rafinesque's scheme for classification was both groundbreaking and deeply flawed. His attempts to assign descriptive names to each of the sightings categories seem clumsy, not to mention highly inconsistent with modern zoology, although one admits that it is easy to suggest this with hindsight.

Even though attempts had been made at detailed, scientific descriptions of sea serpents prior to Rafinesque (for example Guillaume Rondelet's study of the "cetacean centipede" in 1554), no-one had attempted to classify multiple varieties of sea serpent with the same rigor as Rafinesque.

[†] **Baron Georges Léopold Chrétien Frédéric Dagobert Cuvier** (August 23, 1769–May 13, 1832) was a French naturalist and zoologist. He was the elder brother of Frédéric Cuvier (1773–1838), also a naturalist. He was a major figure in scientific circles in Paris during the early 19th century, and was instrumental in establishing the fields of comparative anatomy and paleontology by comparing living animals with fossils. He is well known for establishing that extinction was a fact, being the most influential proponent of catastrophism in geology in the early 19th century, and opposing early evolutionary theories. His most famous work is the *Règne animal distribué d'après son organisation* (1817; translated into English as *The Animal Kingdom*). He died in Paris of cholera.

In 1821, Cuvier made what has been called his "Rash Dictum": he remarked that it was unlikely that any large animal remained undiscovered. Many such discoveries have been made since Cuvier's statement. **From Wikipedia, the free encyclopedia**

[††] **Constantine Samuel Rafinesque-Schmaltz**, as he is known in Europe, (October 22, 1783-September 18, 1840) was a nineteenth-century polymath who led a chaotic life. Many have called him a genius, but he was also an eccentric autodidact, sometimes considered close to insanity. He was very successful in various fields of knowledge; zoologist, botanist, malacologist, meteorologist, writer, evolutionist, polyglot, translator. He wrote prolifically on such diverse topics as anthropology, biology, geology, and linguistics; but was honoured in none during his lifetime. Today, it is generally recognized that he was far ahead of his time in many fields. **From Wikipedia, the free encyclopedia**

Name	Characteristics/Identity
Megophias monstrosus	100 feet long, 2-foot diameter snake-like sea serpent native to the coastal waters of New England. Scaly head and body, sharp teeth, fast swimmer. Possible ally of the sea snake genus *Pelmais*.
Octipos bicolor	58 feet long body, 2-feet long horse-like head, 15-inch wide mouth, long necked sea serpent. Based on a single sighting made off of Irish coast in 1818. Possible Synbranchid related to S*phegebranchus*.
Octipos (?) coccineus	40 feet long, crimson in colour possessing an "acute" head. Based on a single mid-Atlantic sighting. Tentatively placed into the genus *Octipos*.
Pelamis chloronotis	200 feet long sea serpent native to the coast of Newfoundland. Alleged to generate considerable noise, possesses "flexuous hillocks" and a green colouration to its back.

The first rigorous sea serpent classification scheme – The Rafinesque model

Even though the identity conclusions to which he comes, are as implausible today as they would have been in his own time, his work is clearly prototypic of Heuvelmans and other subsequent researchers efforts to build a classification scheme for sea serpents.

1.3 Nineteenth century enthusiasm: Oudemans' *Megophias megophias*

In the first half of the nineteenth century it was common for zoologists to be highly sceptical of sea serpent and lake monster stories that couldn't be dismissed as either hoaxes, or sightings of whales or other large, known animals, but then in 1861 the French gunboat *Alecton* secured a piece of the fabled giant squid (*Architeuthis*), a very old sea monster than can probably trace its pedigree back to Norwegian legends of the *kraken*. From this point on, popular perception changed in the scientific communities of the nineteenth and early twentieth centuries, and zoologists became slightly less hostile to the idea that there could exist large, unknown animals in the oceans and deep lakes of the world. A product of this nineteenth century 'age of enlightenment' was the Dutch Biologist Anthonid Cornelis Oudemans, who produced the first comprehensive overview of sea monster phenomena in his 1892 book *The Great Sea Serpent*. He hypothesized that the majority of sea monster sightings could be explained with reference to a 20 to 200 foot, cosmopolitan, long-necked, previously undiscovered species of pinniped, which he called *Megophias megophias* (literally 'giant snake'). †

Oudemans rejected the idea that there existed multiple sea serpent types in favour of his *Megophias*, which was intended as a distillation of many sea serpent characteristics culminated from many sightings. Oudemans did manage to make a much more convincing case for his single category of sea serpent than Rafinesque did for any one of *his* proposed categories, partly because it is easier to defend a single extraordinary claim than to defend multiple, and partly because he manages to incorporate a considerable amount of sightings data into his monograph. However the idea that there is simply one type of sea serpent behind all of the sightings is unrealistic, especially given the degree to which Oudemans has to 'adjust' obviously incompatible sightings

---

† **Anthonid (Antoon) Cornelis Oudemans** (Nov 12, 1858, Batavia - Jan 14, 1943, Arnhem) was a Dutch zoologist. He wrote his dissertation on flatworms, and in 1885, was appointed director of the Royal Zoological Gardens at The Hague. 1892 saw the publication of Oudeman's *The Great Sea Serpent*, a study of the many sea serpent reports from the world's oceans. Oudemans concluded that such creatures might be a previously unknown large seal, which he dubbed *Megophias megophias*. Reception of the volume has been described as respectful but "cold". Bernard Heuvelmans later suggested that *The Great Sea Serpent* was the root of cryptozoology.

In 1895, Oudemans left The Hague to teach biology in the city of Sneek. **From Wikipedia, the free encyclopedia**

data so that they fit his single category. His book was described as having received a respectful reception within the naturalist community, albeit an unenthusiastic one; evidence that the ultra sceptical 'ice' of Rafinesque's generation of nineteenth century naturalists had thawed a little. Perhaps most importantly, Oudemans' work represents the first serious effort to give the study of unknown animals a scientific gloss, and is considered by many to be a founding document of Cryptozoology.

1.4 Everything changes: The Heuvelmans model

The publication of the 1968 book, *In the Wake of the Sea-Serpents* by the Belgian Zoologist and 'Father of Cryptozoology', Bernard Heuvelmans [†] represents a watershed event in the history of the field. For the first time scientific rigour and a contemporary knowledge of zoology were combined to create a comprehensive account of the sea serpent phenomena.

In effect Heuvelmans revived the Rafinesque assumption of multiple sea serpent identities as a means of negating the contradictions that arose from Oudemans attempt to fit all sightings data into a single category. Heuvelmans re-evaluated the sea serpent phenomena by exhaustively cross referencing hundreds of sightings from which he managed to infer the existence of ten distinct categories of sea serpent [1] into which each and every sighting that was considered to be reliable could theoretically be placed. These categories were built on the premise that certain sightings would correlate with other sightings both in terms of what was being described and location. The development of these categories and the criteria used to assess the reliability of sightings, therefore represented a far more rigorous approach to the study of hidden animals than had ever previously been attempted. Heuvelmans' books are rightly regarded to be amongst the most rigorous on Cryptozoology ever written.

[†] **Bernard Heuvelmans** (October 10, 1916 – August 22, 2001) was a scientist, explorer, researcher, and a writer probably best known as a founder of cryptozoology. His monumental 1958 book, *On the Track of Unknown Animals* (originally published in French in 1955 as *Sur la Piste des Bêtes Ignorées*) is often regarded as one of the best and most influential cryptozoological works. Heuvelmans was born in Le Havre, France and raised in Belgium, and earned a doctorate in zoology from the Free University of Brussels (now split into the Université Libre de Bruxelles and the Vrije Universiteit Brussel). His doctoral dissertation concerned the teeth of the aardvark, which had previously defied classification. Though earlier interested in zoological oddities, he credits a 1948 *Saturday Evening Post* article, "There Could be Dinosaurs", by Ivan T. Sanderson, with inspiring a determined interest in unknown animals. Sanderson discussed the possibility of dinosaurs surviving in remote corners of the world. In 1975 Heuvelmans established the Center for Cryptozoology in France, where his library is housed. In 1982 he helped to found the International Society for Cryptozoology, and served as its first president. He was also the first president of the Centre for Fortean Zoology. **From Wikipedia, the free encyclopedia**

1. Although Heuvelmans derived ten categories in total, all of the categories were never in use simultaneously, as some were eventually dropped from the scheme such as the 'Father-of-all-the-turtles' category and the 'yellow belly' category and a new one was added later in the form of the 'giant invertebrates' category.

Name	Characteristics/Identity
Long necked sea serpent (*Megalotaria longicollis*)	15 to 60 feet long, cosmopolitan, long necked pinniped. Probably an otariid (sea lion).
Merhorse (*Halshippus olai-magni*)	15 to 100 feet long, cosmopolitan, maned, longish necked, large eyed, horse headed pinniped.
Many humped sea serpent (*Plurigibbosus novae-angliae*)	60 to 100 feet long, medium necked archaeocete. North Atlantic in range. Possesses many fixed 'humps' on its back.
Super-otter (*Hyperhydra egedei*)	65 to 100 feet long, medium necked primitive archaeocete or possible sirenian resembling a large otter. Found North of Norway and south of Greenland, presumed extinct, as no sightings have been made since 1848.
Many finned sea serpent (*Cetioscolpenda aelani*)	60 to 70 feet long, armoured archaeocete. Its range is equatorial and subtropical.
Super-eel	30 to 100 feet long eel-like cryptids, cosmopolitan distribution. Multiple eel, synbranchid and elasmobranch identities possible.
Marine saurian	50 to 60 feet long marine reptile with a primarily equatorial distribution. Possible identities include thalattosuchian crocodiles and mosasaurs.
Yellow belly	60 to 100 feet long yellow and black striped tadpole shaped fish.
Father-of-all-the-turtles	A large turtle-like reptile. Has been seen around Newfoundland, Bombay and Soay.
Giant invertebrates	Possible new species or enlarged forms of salp colonial strings and Venus' girdle.

Everything changes – The Heuvelmans model.

Name	Characteristics/Identity
Classic sea serpent	100 feet long, quadrupedal, many humped animal. A conglomeration of Heuvelmans' super-eel, super-otter and many humped sea serpent categories.
Waterhorse	A long necked pinniped showing dimorphism with males possessing manes and females – nasal snorkels. Can transition between both salt and freshwater environments. A conglomeration of Heuvelmans' long necked seal and merhorse categories.
Mystery cetacean	A general category for unusual cetaceans such as unknown beaked whales, double dorsal finned whales and dolphins and high dorsal finned sperm whales.
Mystery saurian	Large, crocodile-like marine reptile. Synonymous with Heuvelmans' marine saurian category.
Mystery sirenian	A general category for surviving Steller's sea cows and out of place dugongs, such as the Saint Helena sirenian.
Mystery manta	A manta species with white dorsal markings.
Great sea centipede	An archaeocete with flexible, retractile fins. Synonymous with Heuvelmans' many finned sea serpent category.
Giant shark	A general category for oversized elasmobranches, which includes surviving megalodons.
Giant octopus (*Octopus giganteus*)	A giant cephalopod native to the tropical Atlantic.
Cryptic chelonian	A large marine turtle, synonymous with Heuvelmans' Father-of-all-the-turtles category.

Post-Heuvelmans – The Coleman & Huyghe model

1.5 Post-Heuvelmans:
The Coleman & Huyghe model

Loren Coleman and Patrick Huyghe developed a classification scheme that follows that of Heuvelmans, and employs a similar inclusion methodology. It was published in their 2003 *Field Guide to Lake Monsters, Sea Serpents, and Other Mystery Denizens of the Deep.*†

Coleman & Huyghe build on Heuvelmans' framework, and have developed categories for cryptid classes that Heuvelmans seems to have overlooked (such as giant sharks and mystery cetaceans), which - with hindsight - Heuvelmans should probably have worked into his scheme. They also include the giant octopus, which Heuvelmans *does* describe in some of his writings, but chooses to keep separate from his sea serpent categories (due perhaps to its distinctly non-serpentine nature).

In some respects Coleman & Huyghe can be described as having taken a more holistic approach to categorizing sea serpents, as many of their categories are general in nature, which has the advantage of making them more flexible in the context of the vagiaries inherent in sightings of sea-serpents and lake monsters.

1.6 The future of aquatic cryptozoology:
The Champagne model

Researcher Bruce Champagne has developed a highly innovative and novel classification scheme for marine cryptids. The Champagne model seems to have developed primarily in response to the methodological criticisms of two authors; Gary Mangiacopra, who in 1992 (and again in 2000) suggested in a series of articles that Cryptzoologists should consider ecological factors in order to accurately ascertain the zoological class of a particular cryptid, and Yasushi Kojo who in 1992 suggested in an article in the now defunct journal *Cryptozoology* that behavioural observations were essential for inferring accurate identities.

---

† **Loren Coleman**, MSW, is an author of books on wide-ranging topics including sociology and cryptozoology. Coleman was educated in anthropology and zoology at Southern Illinois University in Carbondale, and psychiatric social work at the Simmons College School of Social Work in Boston. He did post-masters work in anthropology at Brandeis University and studied sociology at the University of New Hampshire. Coleman taught at New England universities from 1980 to 2004, also having been a senior researcher at the Edmund S. Muskie School of Public Policy from 1983 to 1996, before retiring from teaching to write, lecture, and consult on his many interests.

He has written many books, including a number with Patrick Huyghe, the editor of *The Anomalist*. **From Wikipedia, the free encyclopedia**

Champagne essentially retains the categories developed by Heuvelmans as 'archetypes', but in addition, he employs a multifactor, intensely data-driven methodology to build his own categories or 'types'. Champagne's identity index will not be presented here, as firstly the model is the whole-subject of a chapter entitled *A classification system for large, unidentified marine animals based on the examination of reported observations,* which features in Craig Heinselman's 2007 *Elementum Bestia*, and secondly the index is very large, containing many more identity classes than were proposed by Heuvelmans or Coleman & Huyghe. To present it in its entirety would be tangential to the object of this chapter.

2. COMPARATIVE METHODOLOGIES

2.1 Overview

Heuvelmans' development of a viable, data-driven methodological technique for the synthesis of sightings, and the subsequent extraction of cryptid identities can rightly be considered a significant contributing factor to the establishment of cryptozoology as a scientific field with the potential to make testable predictions. However, already this methodology, whose evolution can be traced back to Rafinesque, is beginning to unravel in the face of criticisms such as those of Mangiacopra and Kojo.

Attempts have already been made to developed a viable alternative to the methodology of Heuvelmans (which this author has termed the 'integrative methodology') and these attempts have taken two forms; there is what could be termed the 'multifactor methodology' of Champagne, and there is the 'plausibility methodology' that this author developed initially for use with data-poor cryptids (see this author's article *Towards a possible caudata identity for the Mongolian deathworm: introducing the 'plausibility method' for identity theory formation amongst lesser known cryptids* in the *2008 CFZ Yearbook*).

2.2 The integrative methodology

The integrative methodology can be described as a data-driven approach to inferring cryptid identities, where identities are extracted from the correlations that can be observed primarily in terms of witness reports and sighting location.

The quality control of such highly subjective witness data is all-important to this methodology, and to that end a large emphasis is placed on factors such as character testimonials and the background histories of witnesses.

The methodology is described as 'integrative' as the correlating factors in the data are combined when inferring cryptid identities, which can be specific with respect to a single species or can represent a potential suite of species.

2.2.1 Ethnozoological and Zoomythological evidence

Ethnozoology is the study of animals largely from the perspective of non-western cultures, whereas Zoomythology is the study of animals depicted in legends and mythology. Heuvelmans, who essentially envisaged Cryptozoology as an interdisciplinary endeavour, suggested that both were potentially valuable sources of data on cryptids,

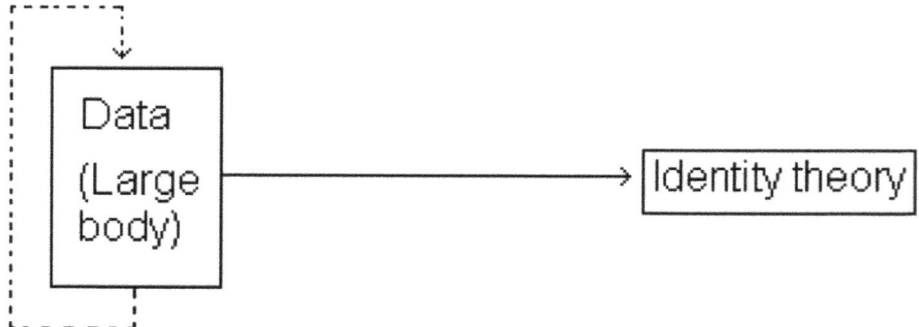

Figure 1: A heuristic diagram showing the workings of the integrative method.

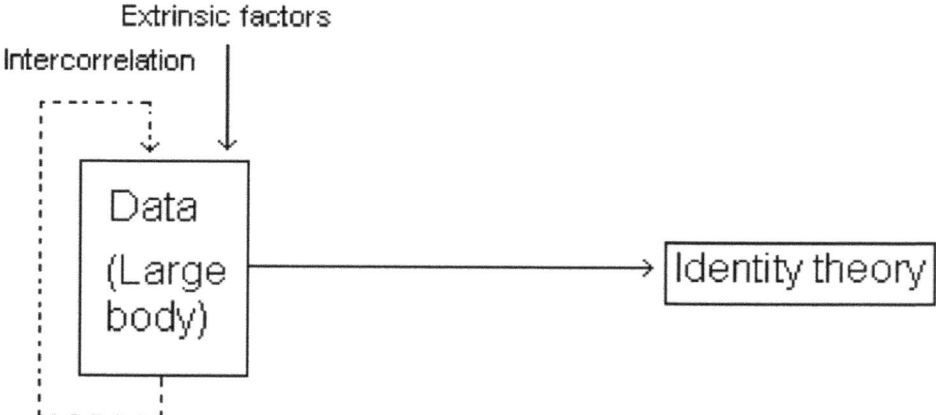

Figure 2: A heuristic diagram showing the workings of the multifactor method.

which although non-confirmatory in their own right, could be used in comparisons with observational data as yet another point of correlation.

Heuvelmans' insistence on the use of folklore and mythology as a 'special class' of observational data has, however, been subject to much in the way of criticism.

Folkloric and mythological accounts of animals, by the very nature of how they have been generated and propagated, will have been affected by the worst kinds of distortion. This has caused some cryptozoologists to simply reject these kinds of 'evidence' on the grounds of their being too subjective to be of any substantial value.

2.3 The multifactor methodology

This methodology which can be said to undergrid Champagne's classification scheme differs from the integrative method in that elements of environment, habitat, and behavioural ecology, are factored into the creation of identity theory indices as additional intercorrelating factors associated with sightings data (beyond simple location and general characteristics), all of which are given a score for reliability.

However, this methodology is still highly data-driven; the main difference with Heuvelman's model being where the bar is set, in terms of the degree of detail that needs to complement a sighting, in order for it to be considered of high quality or not.

2.4 The plausibility methodology

This methodology was developed with a view to evaluating identity theories amongst lesser-known cryptids such as the Mongolian deathworm, for which there exists a paucity of sightings data.

This methodology is hypothesis-driven as it aims to establish tentative identities for cryptids based primarily upon the consideration of extrinsic factors, such as inferred potential niches, and possible evolutionary narratives, in the context of small amounts of available data.

`Occam's Razor` can be used to decide between multiple suggested identities for a single cryptid, leaving the most plausible identity. This identity can then either be falsified and rejected, or refined through the addition of new data.

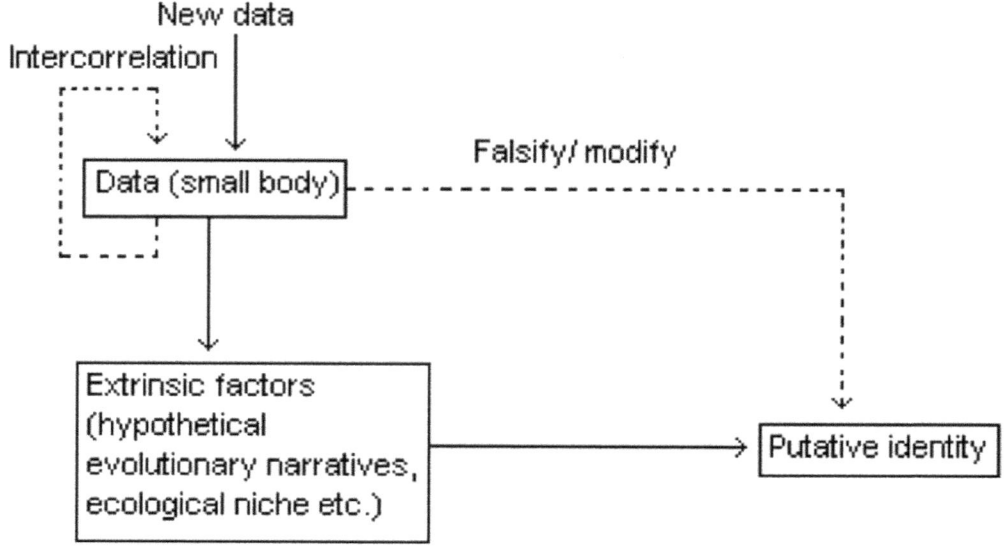

Figure 3: A heuristic diagram showing the workings of the plausibility method.

3. CONCLUSIONS

3.1 Comparisons

The integrative method typically involves taking a body of supporting data and using it to build identity theories by extracting correlations from that data. The strength of a particular cryptid identity theory is therefore simply a function of a) the amount of reliable data available and b) the number of significant points of correlation within the data. Although Heuvelmans can be said to have 'perfected' the integrative method for use in his works, as has been mentioned, the idea of using the intercorrelation of points of agreement in sightings as a means of deriving a taxonomy of cryptid identities, can certainly be seen in the work of Oudemans and to a lesser degree in that of Rafinesque also. The multifactor method of Champagne can be said to be similar, except that what is required of the witness data is a more wide ranging set of extrinsic factors beyond simply location and general descriptions.

The plausibility method is in many respects the inverse of Heuvelmans' integrative method, as it *starts* with the development of an identity theory for a particular cryptid, which is based primarily upon ecological inference and the suggestion of plausible evolutionary narratives. The identity theory is then compared against what data may exist, and re-evaluated in the context of new data via a kind of dynamic feedback that ultimately aims to correlate the extrinsic factors (ecological, behavioural, evolutionary etc.) that went into formulating the identity theory with intrinsic factors (sightings data etc.).

In order to establish an identity theory using the plausibility method, very little data need be known initially - perhaps only enough to establish the geographic location of the cryptid, and some basic facts about its purported morphology etc. `Occam's Razor` plays probably the biggest part in the initial identity determination, as the identity should be plausible above all else, until new data can allow for that identity theory to be revised of falsified.

Both the integrative and multifactor methods, as has been mentioned, typically require large volumes of data to be present. The plausibility method, however, would seem to be much more flexible for assessing less well known cryptids, as putative identity theories can be constructed, and new research directions suggested, using only inference.

Data-driven methodological frameworks, where the quality of the data is generally very low, suffer from two main drawbacks.

Firstly, there is the temptation to commit what could be termed the 'Oudemans Fallacy', namely the tendency to subconsciously interpolate sightings data, so that it compliments a particular category - or categories - of cryptid that the researcher may have preconceived. For example, twenty witness reports could broadly describe a creature with a long neck, four flippers, and a tail, whereas five may describe something with a long neck, four flippers, and no tail.

It is tempting to simply dismiss the five tailless reports, by suggesting that tails were in fact present but not noted by the witnesses, and on that basis suggest that there is only one category of four flippered, long necked, tailed sea serpent where there could just as easily have been a tailless category also. Heuvelmans gets around this by re-invoking multiple sea serpent categories, but still occasionally succumbs to the temptation to lump sightings in this manner.

The plausibility method overcomes this by employing primarily extrinsic factors to determine identity. Going back to the hypothetical data set; if it were found, for example, that the tailless long-necks were seen predominately in the northern hemisphere, whereas the tailed ones were only seen in the southern hemisphere then it would be reasonable to speculate two plausible identities on the strength of there existing two similar species possessing different ranges so as to avoid competition.

The second criticism of data driven methodologies is primarily aimed at the multifactor method. Even when good quality filters are employed (background checks on witnesses, corroboration etc.) the data are still highly subjective, so interpolating all of the additional extrinsic variables associated with a sighting (environmental, temporal, behavioural etc.) into the formation of an identity theory, simply invites the possibility of making false correlations.

Although the plausibility method for inferring cryptid identities is not strongly-data driven, this does not mean that it rejects the need to acquire data. It is simply that cryptozoology should be a primarily hypothesis-driven science, where the subjective limitations of the available data should always be carefully considered by researchers.

3.2 And finally, the object of the manuscript

The purpose of this manuscript is to use a combination of contemporary zoological knowledge and aspects of the plausibility method to re-evaluate Heuvelmans' final eight cryptid identities, so as to shed light on possible new and alternative identity theories - something that is in many respects long overdue.

Although this author does not believe Heuvelmans' categories to be entirely representative of all that there is in the way of undiscovered, large bodied sea creatures; due to the meticulous manner in which they were derived and the abundance of various data ideally suited for novel analysis and interpretation, Heuvelmans' proposed identities form a good and solid basis for the proposed analysis.

CHAPTER 1.
Heuvelmans' Pinnipeds

· The long necked sea serpent

· The merhorse

Speculations on the Long Necked Seal theory.

1. INTRODUCTION

1.1 Overview

In this chapter, a case will be built for a pinniped identity for Heuvelmans' long necked and Merhorse categories, not just using sightings data, but a detailed evolutionary ecological analysis also. Heuvelmans' pinnipeds are unique in this respect, when compared to his other proposed categories, as there exists a considerable volume of sightings data from which potential ecological and evolutionary relationships can be inferred.

Each chapter in this monograph is designed to be a detailed stand-alone survey of a particular class of Heuvelmans-derived sea serpent. Separating the various sea serpent categories into different chapters represents an attempt to provide the sort of rigorous focus and specialization that cryptozoology has often been accused of lacking by its mainstream detractors.

1.2 Identity theories

Historically a variety of theories have been put forward to try and establish the identity of this long necked aquatic cryptid – perhaps the most endearing single image that comes to peoples' minds when they think of lake monsters or sea serpents. These identity theories include:

1.2.1 The 'Elasmosaurus' theory

The theory that longnecks are 'living fossil' elasmosaurs or other long necked marine reptiles that somehow survived the KT extinction event at the end of the Cretaceous is an old one. It probably dates back to the discovery of the first plesiosaur fossils by Mary Anning in the 1820s.

It is the most popular and well-known theory today regarding the identity of aquatic cryptids. Advocates argue that the meteorite that facilitated the KT extinction may have had less of an impact on marine ecosystems than on terrestrial ones, therefore if remnant populations of large marine reptiles survived in remote and isolated stretches of water, they could have continued existing and evolving to the present day.

Figure 1. A plesiosaur. By Adam Stuart Smith (2002).

1.2.2 The Basilosaurus theory

Zoologist Roy P. Mackal advanced this hypothesis as a more 'realistic' alternative to the elasmosaurus theory. It holds that long necked sea serpents and lake monsters are the result of modern day basilosaurs, or zeuglodons (as they are also called), which have survived extinction 37 million years ago. Early reconstructions, such as that of Albert Koch in 1845 attributed to the zeuglodon a long serpentine neck and body, also it was suggested that because zeuglodons are mammalian cetaceans, they could exist across a range of environmental gradients such as temperature and salinity that might pose an obstacle to the reptilian elasmosaurs and their allies.

1.2.3 The Giant squid theory

This theory, which was first advanced by Henry Lee in his 1883 book *Sea Monsters Unmasked*, proposed that longneck sightings were caused by giant squids near the surface whose tentacles when raised out of the water could have been mistaken for necks. Its suggestion came after *Architeuthis* gained acceptance within the naturalist community of the nineteenth century, and for a while became the definitive explanation for all sea serpent sightings.

Figure 2. A giant squid 'posing' as a sea serpent. From Lee (1883).

However, there existed little consensus as to why *Architeuthis* would raise its tentacles out of the water. Theories ranged from the idea that it was testing the wind direction, to the idea that it was deliberately mimicking something more menacing in order to deter potential predators whilst near the surface.

1.2.4 The long necked seal theory

This theory, in its contemporary form, suggests that an approximately 15 to 100 foot long, previously undiscovered, elasmosaur-like type of pinniped might be the culprit behind many sea serpent and lake monster sightings. As has already been mentioned, this theory can be traced back to Oudemans, who proposed the existence of *Megophias megophias*, a hypothetical 20-200 foot long pinniped that was theorized to be a representative of a hypothetical transitional form between whales and seals.

Heuvelmans refined the dimensions of Oudemans' super-seal and suggested that it was an otariid - a relative of the sealion. He also suggested that there might in fact be two pinniped species behind longneck sightings; the 'true' long necked seal *Megalotaria longicollis* (literally 'the big sealion with a long neck') and *Halshippus olai-magni* (literally 'the sea horse of Olaus Magnus' - named after the Swedish ecclesiastic who first described it in writing), [2] which he called the Merhorse. Unlike *Megalotaria*, *Halshippus* is believed to have a shorter length neck, larger eyes (as it is supposed to live at greater depth), a distinctive horse or camel shaped head, and is a little longer overall (30-100 feet vs. 15-60 feet for *Megalotaria*). Heuvelmans regarded both species as being highly cosmopolitan in their distributions, although sightings of both species are more frequently made in the northern hemisphere.[3]

The long-necked seal theory has never enjoyed the same popularity as the elasmosaur theory but has on occasion featured in prominent cryptozoological literature such as Peter Costello's *In Search of Lake Monsters*.

As a point of reference, the term 'longneck' shall be used in describing both genera of this speculative pinniped throughout this chapter. When distinctions are to be made between the two, the pinnipeds shall be referred to by their separate genus names.[4]

There have, of course, been other theories put forward to try and explain away longneck sightings. Heuvelmans suggests that some sightings could be explained by super-eels surfacing.

---

2.   It should be noted that these parataxa names given by Heuvelmans to his various types of sea serpent can not be truly valid, as for a species to be *formally* identified with such a name there needs to exist a type specimen. However the names are descriptive, creative, and remove the ambiguities associated with the creature's vernacular names, so they will continue to be used throughout this chapter.
3.   This is, of course, not to suggest that both species are less abundant in the southern hemisphere, only that they are *observed* more frequently in the northern hemisphere. This disparity could be due to the fact that a larger percentage of the oceans of the northern hemisphere have been, and are, traversed by ships which means that there are fewer opportunities to observe longnecks in the southern hemisphere.
4.   As was mentioned in the introduction Coleman and Huyghe, like Costello, combined *Megalotaria* and *Halshippus* into the single category of 'Waterhorse' where the differences between the two are attributed to sexual dimorphism within a single species. However, the marked morphological and ecological differences between the two would seem to suggest to this author that Heuvelmans was probably correct in classifying them as different species in different genera.

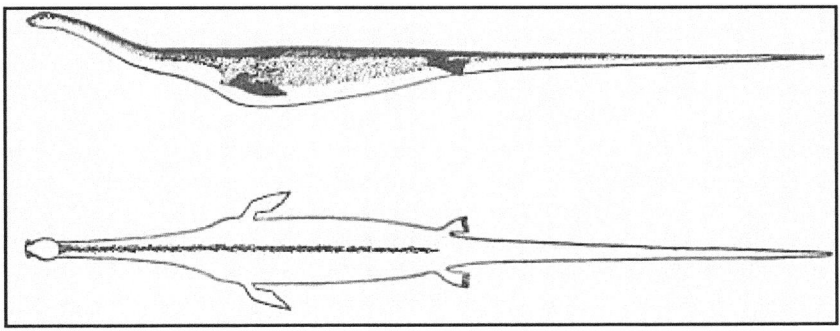

Figure 3. Megophias megophias *from Oudemans (1892).*

Other theories include the idea that they represent previously undiscovered long necked amphibians (as Roy P. Mackal once proposed) or that they are perhaps even serpentine super-otters. It has also been suggested by sceptics that some longneck sightings could be of large snakes such as pythons, or *even* elephants holding their trunks out of the water whilst swimming. These theories warrant no further elaboration here, as - historically - they have not stimulated as much debate in cryptozoological circles as the aforementioned four.

1.3 Which is the more likely explanation?

First of all, it must be noted that this author is making a fairly substantial assumption in discussing the relative merits of these identity theories, as it is being assumed that there is *something* that needs identifying. Of course, as has been mentioned, sceptics will argue that a case can be made for the idea that longneck and sea serpent/lake monster sightings in general can be universally dismissed as hoaxes, or as misidentifications of more mundane species (or objects) and in that respect, the sceptics serve a valuable and welcome function as counters to the sort of over enthusiasm that can distort objectivity. Although evidence will be looked at for the existence of these creatures in subsequent sections, the primary purpose of this monograph is as an exercise in theoretical zoology, and is therefore concerned with speculation albeit of a scientifically informed nature. To that end, this monograph shall not overly concern its self with the more sophisticated sceptical arguments, and if the readership should wish to familiarize them selves with these then they should consult works like folklorists Michel Meurger and Claude Gagnon's *Lake Monster Traditions: A Cross-Cultural Analysis*, which is considered by many to be the best counter to Heuvelmans written to date.

As has already been mentioned, the Elasmosaurus theory of longnecks has by far the most currency within cryptozoological circles but is this necessarily justified? Plesiosaurs, elasmosaurs and their relatives were large carnivorous, viviparous marine reptiles that, contrary to popular belief, were *not* dinosaurs. What would it have taken for them to have survived the K-T extinction event? It would have required marine ecosystems to have been less damaged by the meteorite impact than terrestrial ones. There is indeed some evidence that this was the case; marine turtles for example survived the K-T event, as did the coelacanth; advocates of the Elasmosaurus theory cite both as evidence of the possibility of post K-T survival. It is only logical that some would ask that if turtles and coelacanths could survive the extinction event, then why not elasmosaurs and their relatives? Size and position in the food web would seem to be the main limiting factors in this instance; larger species that were dependent upon other species for food would probably have been far more vulnerable to the environmental changes, and other extinctions facilitated by the meteoritic impact. Let's assume that large marine reptiles from the Cretaceous managed to survive the K-T event; is the reptile identification compatible with the facts concerning longnecks? Lets look at distribution. Both genera of long necked pinniped are described by Heuvelmans as being cosmopolitan; a cold blooded reptile is going to prefer the warmer waters of the tropics to the colder waters closer to the poles. A cosmopolitan distribution is more consistent with longnecks being mammalian rather than reptilian. Furthermore, if it is to be believed that lake monster sightings are the product of latter day elasmosaurs and their ilk, then it has to be concluded that somehow they managed to evolve [5] an adaptable tolerance to *both* fresh and saltwater environments [6]. Mammals are far more likely to be readily tolerant of both salt and fresh water environments. Another thing worth considering are the descriptions of longnecks. They are described as being able to hold their necks high out of the water, which indicates that they are maximally flexible through the vertical plane - again a mammalian trait. Reptiles are primarily flexible through the horizontal plane (snakes, for example, rely on side to side movements for locomotion), and elasmosaurus is now known to have possessed a brittle neck that could not have been held out of the water.

The Basilosaurus theory has to its credit, the fact that zeuglodons were mammalian cetaceans, which could - as was mentioned previously - potentially have survived across a much larger spectrum of environmental gradients than could elasmosaurs. However, the visual descriptions of long-necked sea serpents do not correlate well with what is now known - from modern reconstructions - about the morphology of zeuglodons. It was mentioned previously, that in early reconstructions, zeuglodons were often portrayed with long necks (see figure four). However, in more *recent* reconstructions, the neck has been shortened considerably (see figure five). If a zeuglodon were seen alive today, it would more likely be compared to a large crocodile than to an elasmosaur.

This leaves the giant squid whose tentacles held high out of the water are supposed to resemble the necks and heads of aquatic cryptids; a claim which despite its vintage, and the unlikely circumstances underpinning the squid's motivation to perform this bizarre act, is still repeated today in texts on cryptozoology.

---

[5] As the vertebrate Palaeontologist Darren Naish has observed, the extremely conservative nature of plesiosaur evolution (they retained essentially the same body plan throughout their 160 million years on earth) would seem to make scenarios where plesiosaurs have evolved into an essentially modern mammalian form unrealistic in the extreme.

[6] While it is the case that several fossil plesiosaur specimens have been recovered from what would appear to have been freshwater systems, it is known that many reptiles are not capable of internally regulating against changes in environmental salinity as efficiently as mammals - a good example being the freshwater crocodile of Australia. The plesiosaurs that are recovered from sediments associated with freshwater systems tend to be small and date from the late Jurassic/early Cretaceous boundary. No elasmosaurs have yet been identified as ever having made use of freshwater systems. Based on this it is a reasonable assumption to make that as plesiosaurs became more specialized marine hunters they may well have lost their fresh water tolerance.

Figure 4. *Koch's 1845 'Hydrarchos'. It was in fact assembled from the remains of five specimens that turned out to be both basilosaur and non-basilosaur in nature.*

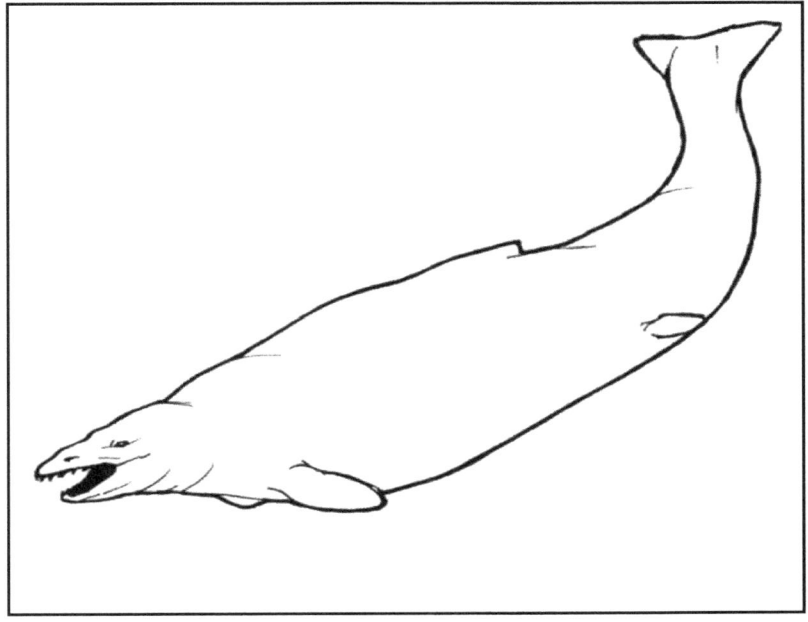

Figure 5. *A modern reconstruction of basilosaurs, note the absence of a long neck.*

Squids are unable to hold their tentacles out of the water for the simple reason that the muscular system that provides their bodies with mechanical support could not produce enough force to sustain the weight of a tentacle out of the water. The giant squid theory runs into the same obstacle as the Elasmosaurus theory when dealing with the question of freshwater habitats. A giant squid could not survive in a freshwater environment, although this fact hasn't prevented advocates of the giant squid hypothesis from speculating that there exist species of freshwater giant squids (for example Tony "Doc" Shiels, who took a famous but controversial photograph of 'Nessie' in 1977, has tongue-in-cheek proposed the existence of a bizarre 'elephant squid' to explain lake monsters).

All the observational and ecological data seem to support the theory that longnecks are mammalian in nature, but what sort of mammal are they? For morphological and palaeontological reasons pinnipeds seem to be the only likely candidate mammals for the longnecks, the case for which will be elaborated in section three. This was Heuvelmans conclusion, and shall form the basic identity assumption underpinning all subsequent discussion in this chapter.

2. AN EVOLUTIONARY HISTORY OF THE PINNIPEDS

2.1 Overview

Now that it has been suggested that longnecks (should they exist) are more than likely mammalian, and very probably pinnipeds. It may be useful at this juncture to become familiar with the evolutionary history and taxonomy of these aquatic mammals. An understanding of seal morphology, behaviour, and ecology, will be necessary before the characteristics of these hypothetical long necked pinnipeds are discussed in the next section.

2.2 Taxonomy

Seals, walruses, sealions, and their allies are collectively referred to as pinnipeds. Pinniped means 'feather feet'. This used to be the designation for their suborder (pinnipedia), however in recent years they have been reassigned to the suborder Caniformia, which currently contains three families:

Odobenidae: This includes the walruses and contains a current total of one species in one genus.

Otariidae: This contains the sealions and fur seals and was the family into which Heuvelmans placed one of his hypothetical long necked pinniped genera – *Megalotaria* (Heuvelmans never explicitly suggests an identity for *Halshippus*, apart from calling it a pinniped, however he does suggest that there are similarities between its fur and skin to that of the sealions). There are currently twelve species in six genera within this family.

Phocidae: Which comprise the 'true' seals and currently contains nineteen species in thirteen genera.

Despite the taxonomic reclassification, seals, walruses, sealions and their allies are still generally referred to as pinnipeds, although this term is interchangeable with 'caniform'. 'Pinniped' shall be used exclusively in this chapter, as it is the more commonly used collective term for creatures in the aforementioned families. The closest living relatives to pinnipeds are the bears, according to both molecular and morphological studies.

2.3 Polyphyletic or Monophyletic?

For a long while it was thought that the phocids (the true seals) derived from an otter-like ancestor whilst the other two families (the Walruses and the eared-seals) derived from a bear-like ancestor, making the pinnipeds polyphyletic (not sharing a common ancestor). Molecular data indicate however that the families of pinniped are all related through a common ancestor, making them monophyletic.

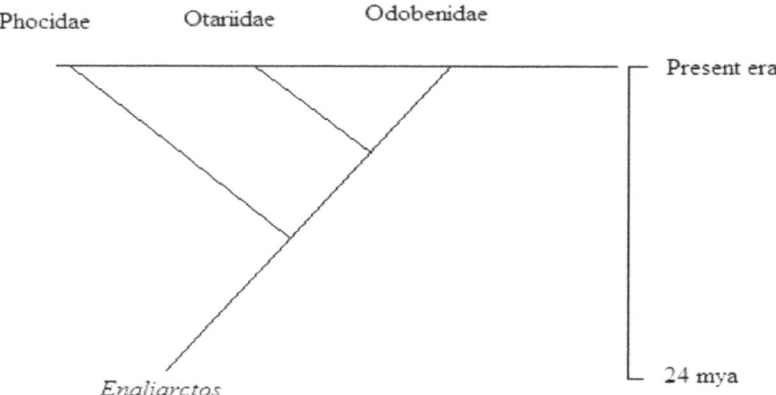

Figure 6. A cladogram showing the evolution of the three families of pinniped.

All pinnipeds are thought to have diverged from a bear-like ancestor approximately twenty-four million years ago during the late Oligocene epoch. The earliest fossil pinniped is the *Enaliarctos*, which lived approximately twenty-two million years ago.

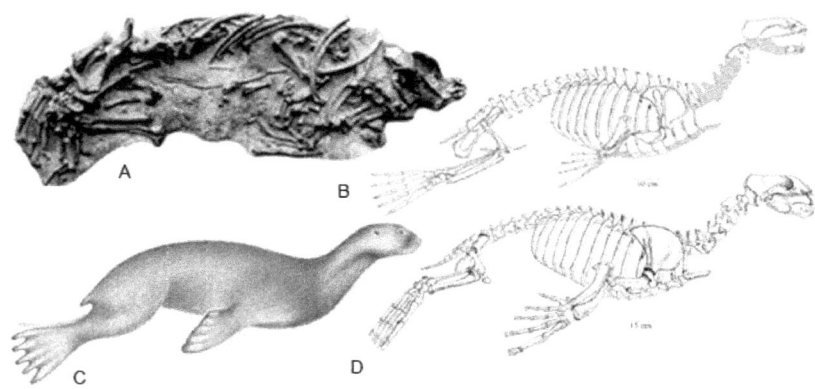

Figure 7. Enaliarctos from From Wikipedia, the free encyclopedia

Enaliarctos was probably adapted to an amphibious lifestyle much like an otter, and unlike modern pinnipeds it seems to have possessed a short tail. It did however also seem to posses some modern sealion-like characteristics, such as a sophisticated inner ear design, that would have been useful underwater for predator/prey sonic detection, and large eyes.

2.4 Ecology

All pinnipeds are carnivorous to varying degrees. Most prey upon crustaceans, krill, shellfish, squid etc. The leopard seal (*Hydruga leptonyx*) is by far the most carnivorous and predatory of the pinnipeds, as in addition to penguins, it is known to occasionally *even* incorporate other seals into its diet.

Pinnipeds tend to occupy positions near the top of their food chains, but not at the very top. Their natural predators can include humans, orcas, polar bears, and - as in the case of the leopard seal - sometimes other pinnipeds. The families are widely distributed, with their members being found at both poles, and in the temperate and tropical zones.

As primary and secondary consumers, pinnipeds play a significant role in regulating their surrounding ecology.

2.5 Reproduction

A curious feature of pinnipeds is their reproduction. They haul themselves onto land, and move a distance away from their feeding grounds, the distance being dependent upon the family. Otariids (sealions), for example, can move a considerable distance over land due to the presence of opposable hind flippers, whereas phocids have to drag themselves using their fore limbs, so can not cover much ground on land.

Due to the fact that pinnipeds have to traverse a distance over land or ice in order to get to their breeding grounds, they exhibit very high degrees of reproductive synchrony, where all the individual members of a particular population become reproductively active within a very short time window.

A breeding colony of pinnipeds is called a rookery. High-status males tend to compete with one another for females, whilst females gather together as a defensive mechanism, especially against harassment by lower-status males.

2.6 Key differences

As will be elaborated upon in the next section, many of the traits exhibited by known pinnipeds could conceivably be compatible with what is known about longnecks; however there are some morphological, behavioural and ecological traits, that would seem to make longnecks quite unique amongst pinnipeds, if indeed this is what they are.

3. THE NATURE OF THE BEAST

3.1 Formally introducing the longnecks

As pinnipeds go, size wise the longneck species described by Heuvelmans and subsequent researchers are big, the very biggest in fact. To illustrate this, the largest known pinniped is the male southern elephant seal (*Mirounga leonine*). It averages out at just over four metres (thirteen feet) in length, and weighs just over two tonnes (2,200 Kg). Heuvelmans estimated that *Megalotaria longicollis* could sometimes reach 65 feet in length, and *Halshippus olai-magni* could occasionally reach over 100. They would probably therefore weigh several tonnes each when fully grown.

Appearance wise, longnecks are quite unlike any other pinniped species. They are characterized by elasmosaur-like necks, with disproportionately small heads. People who have reported seeing them have described them as having 'humps' (caused by the presence of back fat deposits and the vertical plane undulation of the body), dark brownish fur which sometimes contrasts with a lighter coloured underside fur (primarily a characteristic of *Megalotaria*) and occasionally a main of reddish neck hair (primarily a characteristic of *Halshippus* which may be a sexually selected trait). The eyes are variously described as forward pointing and large, with a reddish hue (in the case of *Halshippus*) or small, black, and practically invisible (in the case of *Megalotaria*). Another interesting morphological characteristic of longnecks is the presence of two horn-like protrusions from the tops of their heads - a feature most prominent in *Megalotaria,* although it is occasionally seen in *Halshippus*.

Peter Costello has implied in his writings that these structures are *pinnae* (cartilaginous projections that form the external ear). This makes an excellent case for longnecks being placed in the otariidae family, as Heuvelmans suggested - although not for this particular reason. Heuvelmans suggested that these horn-like structures, instead of being pinnae, may in fact be erectile tubules that allow the pinnipeds to obtain air without ever having to fully surface. His colleague, the noted Cryptozoologist Ivan T Sanderson, suggested another function for them, namely that they could provide a useful means of venting waste respiratory gasses in such a way as to prevent the bubbles from obscuring the pinniped's field of vision underwater.

As far as biological snorkels are concerned [7], no analogous structures exist in other pin nipeds, so the case is weakened for this, given that such a snorkel system would require a completely novel and unprecedented re-ducting of the pinniped nasopharyngeal system. Additionally, no plausible evolutionary scenario can be arrived at to explain the adaptive benefit of having these. Surely a pinniped of these dimensions would have little to worry about in terms of surface predation? Additionally, the majority of air is likely to be expelled at or near to the

---

7   Of course there are examples of specialised nasopharyngeal adaptations in the pinnipeds, but these take the form of inflating sacks that do not function as biological snorkels. Examples of this form of adaptation include the proboscis of the male elephant seal, which is used for moisture conservation and vocal amplification, and the 'bulge' of the male hooded seal, which serves a display purpose.

surface prior to diving, rather than throughout the dive, which means that they would provide minimal benefit as vision aids.

The bottom line is that neither Heuvelmans' nor Sanderson's proposed functions for these structures seem plausible.

The horns, as have been mentioned, are reminiscent of pinnae. In longnecks it could be the case that the ears have developed in a special manner so as to resemble horns. There appear to be variations between individual sightings as to the lengths and prominence of these horns; perhaps this is evidence of dimorphism, indicating the presence of a sexually selected trait, or perhaps there is considerable variation within a single species, or between species of longneck, as certain reliable sightings - and even photographs - of what may in all likelihood be long-necks, do not indicate the presence of these 'horns' at all.

Other dimorphic traits may include variations in head shape and length (which may also vary with age), the presence of a main of hair, neck length, and body size.

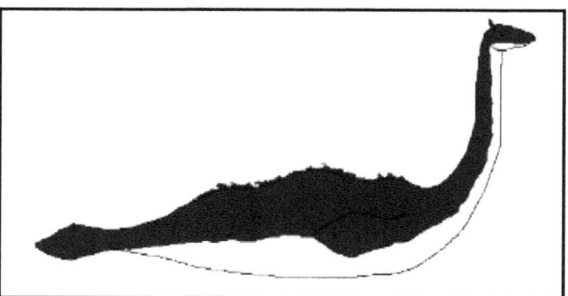

Figure 8. *Megalotaria longicollis* by Cameron McCormick (2006).

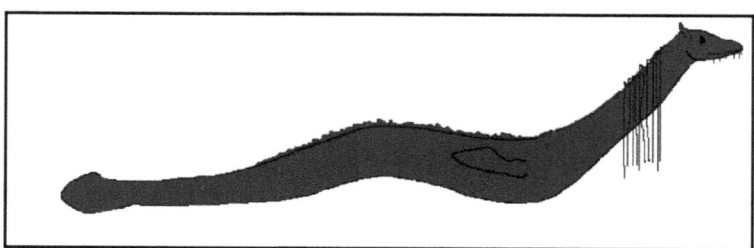

Figure 9. *Halshippus olai-magni* by Cameron McCormick (2006).

3.2 Convergent evolution

A fundamental thesis of this chapter is that the various species of long-necked pinniped, if they exist, have convergently evolved to fill similar ecological niches to the ones left vacant by the extinction of elasmosaurs and their allies, and that any rigorous feasibility analysis should start with an examination of the properties of those niches.

The ecological niche can be loosely defined as the sum of all the resources that a species needs to exist. Each niche is unique, as species will differ from one another in terms of their particular resource requirements. However two genetically distinct - but contemporary species - can occupy very similar niches. A good example of this can be found in the form of the New World vultures and the Old World vultures. Both are extremely similar to one another in terms of appearance and feeding habits, and for a long time they were considered to be closely related. It took the development of molecular genetic techniques to determine just how unrelated these two types of bird really were. Whilst the Old World birds are raptors, their transatlantic analogues are closely related to the storks!

When different species evolve to fill similar niches separated by long periods of time, the process is known as `evolutionary relay`. A good example of this can be found in the comparison between extinct members of the reptilian order ichthyosauria, and both extinct and extant members of the mammalian order cetacea; both orders have produced species that are very similar to one another morphologically, and have clearly adapted to exploit very similar niches. As a matter of fact the existence of members of the cetacea as what are effectively modern day mammalian ichthyosaurs, seems to lend added credence to the possibility of a pinniped becoming a modern day mammalian elasmosaur, as it indicates that there exists both a mechanism, *and* a precedent for such an evolutionary event. [8]

3.2.1 The niche of the elasmosaurus

Aside from the morphological similarities between the elasmosaurs and the longnecks (both are large in size - occupying roughly the same size range, both have long necks, four flippers etc), if the assumption that the longnecks occupy a similar niche to elasmosaurs is accurate, then there should also exist modern day analogues to the resources that elasmosaurus had available to it in the Cretaceous marine environment.

Studies of fossils indicate that elasmosaurus probably preyed extensively upon a particular species of fish called *Enchodus* that grew to about one to two metres in length. Is there an ana-

---

[8] Naish has suggested that long necked marine mammals would suffer from the drawback of not being able to adequately maintain their core temperature at their heads, due to the distance between the head and body. This morphological constraint would certainly seem to pose a challenge to the feasibility of very long necked pinnipeds; however it would be interesting to see how the swan-necked seal, known only from fossils, compensated for this. One solution could be that perhaps the arteries feeding the head with blood evolved to be located deep in the neck, and were therefore able to better take advantage of the insulating properties of blubber.

logue to this fish in modern marine ecosystems? Various cod species including the Atlantic cod are known to be able to reach two metres in length, so in this respect at least, they are like *Enchodus*. An interesting point to note is that according to Heuvelmans, the majority of *Megalotaria* sightings take place around the UK coast and to a lesser degree around the New England coast, indicating that their distribution is primarily Atlantic, which also seems to correlate geographically to a degree with the distribution of Atlantic cod. Another interesting point to note is that according to the fisheries data, Atlantic cod stocks (measured by annual catch sizes) rose sharply between 1950 and 1955 before declining and then rising again even more sharply after 1965. Heuvelmans suggested that the longnecks were on the rise as a whole, as the frequency of sightings was increasing through the 50s and into the 60s.

Robert Cornes, a Cryptozoologist who has focused his studies on longnecks, and other researchers, suggest that the frequency of sightings has declined in recent decades, and that this could be due in part to the fact that more areas of the world's oceans are being traversed than in previous decades making them less hospitable to the shy longnecks. This is plausible, but it could also be due to the recent sharp decline in cod stocks. If cod is a primary prey item of longnecks, then it stands to reason that the frequency of sightings should track the cod abundance data to a degree.

There should also be evidence that both elasmosaurs and long necked pinnipeds utilized similar habitats, and indeed there is. Paleoecological evidence suggests that elasmosaurs preferred coastal marine habitats. Heuvelmans' sightings distribution data indicate that this is also the case for longnecks, as the vast majority of sightings have taken place near coastal regions.

Another way in which long necked pinnipeds could be similar to the elasmosaurs, is in terms of the behavioural ecology of how they give birth. For a long time it was thought that plesiosaurs, elasmosaurs, and their allies came ashore to lay their eggs, but fossil remains were discovered that indicated vivipary (or the giving birth to live young rather than egg-laying). If the longnecks *are* occupying a similar niche to the elasmosaurs, then it stands to reason that they could have evolved the ability to give birth at sea. This would be fairly unique for a pinniped, as all known pinnipeds come ashore to give birth [9], but it is not unique as far as aquatic mammals in general are concerned; there is a precedent for it in otters. It would also require the offspring to have very short maturation and weaning times. Otariids as a rule have longer weaning and maturation periods. However juveniles of certain phocid species, like the grey seal mature very quickly. It would not take that much of an imaginative leap to reasonably speculate the existence of marine parturition and rapid maturation in longnecks, as there seem to be precedents for both within pinnipeds and other marine mammals.

3.2.2 How longnecks and elasmosaurs seem to differ

Two separate species can occupy similar niches, but can also *differ* from one another markedly. There do seem to be some elements of reported behaviour in longnecks that make them

---

9 Although Heuvelmans informs us that marine parturition has been observed in grey seals, based upon information that he received from the Marine Biologist L. Harrison Matthews.

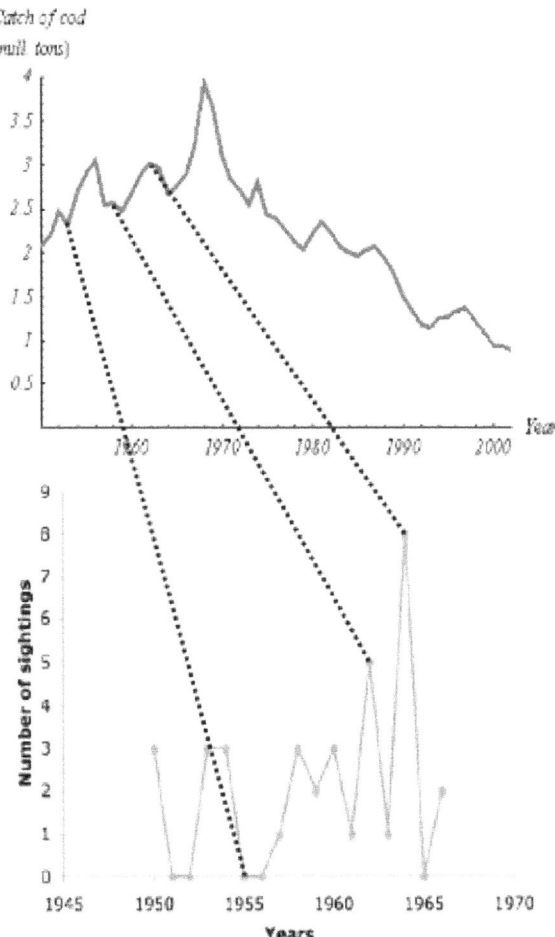

Figures 10 & 11. *The top graph shows the change in the catch size over time for various species of Atlantic cod. The plot represents the total catch sizes for both the Northeast and Northwest Atlantic cod species. The bottom graph shows the increase in longneck sightings (both* Megalotaria *and* Halshippus*) between 1950 and 1966. Note the visual correlation between the two, suggesting that there may be a relationship between cod abundance and longneck sighting frequency. (Data from the Food and Agricultural Organization and Heuvelmans).*

Figure 12. *The swan necked seal reconstructed by Darren Naish (2006).*

quite distinct from their reptilian elasmosaur counterparts.

Firstly, cryptozoologists like Heuvelmans, Costello, Huyghe and Coleman have argued in their writings, that longnecks would seem to be able to make the transition from marine to freshwater habitats, as they are thought to be behind both certain sea serpent and lake monster sightings. This is an advantage that would potentially be afforded them by their mammalian physiology; large marine reptiles, as were discussed in the introduction, would probably be more intolerant of making the switch to freshwater environments.

Longnecks seem also to have a cosmopolitan distribution (unusual by pinniped standards) and have been sighted in both warm and cold-water environments. Large marine reptiles in general would have been prevented from having had a similar distribution, due to their reptilian piokilothermic (cold-blooded) physiology. They would probably have preferred the warmer waters of the tropical seas.

There have been reports of lake monsters in particular being seen on land; something that would be quite impossible for elasmosaurs. Several such sightings have been made in - and around - the Loch Ness area, interestingly enough. If these sightings are to be regarded as sightings of long-necked pinnipeds (and there are some elements of these sightings that might indicate this) then this is evidence that at least one longneck genus (the *Megalotaria*) has the ability to move on land in a reasonably efficient manner, similar to that of the sealions, which would be most useful when translocating from marine to freshwater habitats.

Why longnecks would choose to migrate from one variety of habitat to another completely different one, is difficult to fathom. Perhaps it is the case that longnecks only ever come to deep-water lakes to breed, or maybe it is a response to competition pressures in marine environments. Perhaps even the ability to transition serves the same evolutionary purpose that it does in cetaceans; namely it allows them to be rid of marine parasites that are intolerant of freshwater.

3.2.3 Long necks in other pinnipeds

Longish agile necks seem to be a common trait amongst pinnipeds. As was discussed in the previous section, longish necks are a trait of the otariidae; however there are other pinnipeds that also exhibit longish necks. Leopard seals - for example - have longish necks, which they are capable of using to quickly strike at prey. There is even an extinct species of phocid encountered briefly in footnote eight, called the swan necked seal (*Acrophoca longirostris*), whose neck was very plesiosaur-like, as it appeared to be sinuous as well as long, and probably advantaged the seal when taking prey, as in the case of the leopard seal.

3.3 Migration patterns and habitats

Megalatoria and *Halshippus* seem to have similar patterns of migration. *Megalotaria* appears to spend spring in the Northern temperate regions. During the autumn, it is thought to migrate to the tropics. It has been suggested that it may migrate onwards into the southern hemisphere - pursuing the summer. This pattern of migration would account for the cosmopolitan nature of its distribution. Less can be inferred about the migration patterns of *Halshippus,* except that its cosmopolitan distribution would also seem to indicate that it migrates in a similar manner to *Megalotaria*. It seems to possess adaptations to a deep-water lifestyle, so probably spends some of its time at abyssal depths. This behaviour may vary seasonally with the availability of prey.

3.3.1 Competition between longneck species

The sightings data indicate that there exists a considerable general range overlap between *Halshippus* and *Megalotaria*, suggesting that there may be competition between the two genera; and - indeed - Heuvelmans' sightings data may be interpreted to suggest that such competition exists in certain regions where there is a co-occurrence of the two.

According to the data, the Pacific west-coast of North America would seem to have provided *Halshippus* with an ideal set of niche conditions. The same could be said of the British coastal waters for *Megalotaria*. In the two regions, each respective species seems to have a significant advantage, effectively allowing one to successfully out-compete the other. In the Atlantic Ocean, off the eastern coast of North America, and in the Australasian Pacific region, *Megalotaria* and *Halshippus* can also be said to have respective numerical advantages. However the competition does not seem to be as pronounced as it is in the other regions. This could be put down to one of three reasons; firstly there could simply be less sightings data available for these regions, therefore what Heuvelmans reports, is not representative of the true magnitude of the competition in those areas. Secondly, it could be argued that the ocean around New England and Australia allows *Halshippus* to take advantage of its speculated abyssal adaptations, which would reduce the degree to which it impinges on the niche of *Megalotaria*, thus reducing competition. Thirdly, perhaps the coastal waters of both New England and Australia are not particularly suitable for either species, which would account for the generally low numbers of sightings in those regions.

These inter-specific competition data for longneck genera also have the potential to give valuable insights into how the two genera evolved. *Megalotaria* may represent the common ancestor species, as it possesses characteristics such as the ability for land locomotion, which potentially make it less advanced than *Halshippus* with its hypothesized adaptations for an exclusively aquatic lifestyle. *Halshippus* could have evolved as a result of populations of *Megalotaria*, being forced to adapt to using more abyssal habitats in response to competition nearer the surface, which resulted in the evolution of larger, forward pointing eyes, and a highly streamlined body shape with an attendant loss of the ability to travel on land.

3.3.2 Competition with other species

Heuvelmans suggested that longnecks (particularly *Megalotaria*) might at one point have been in direct competition with another hypothetical long necked cryptid, namely the super-otter (*Hyperhydra egedei*), about which more will be said in the subsequent chapter. Its habitat was originally the North Atlantic according to sightings data collected by Heuvelmans. The decline in super-otter sightings seems to correlate with an increase in longneck sightings (*Megalotaria*

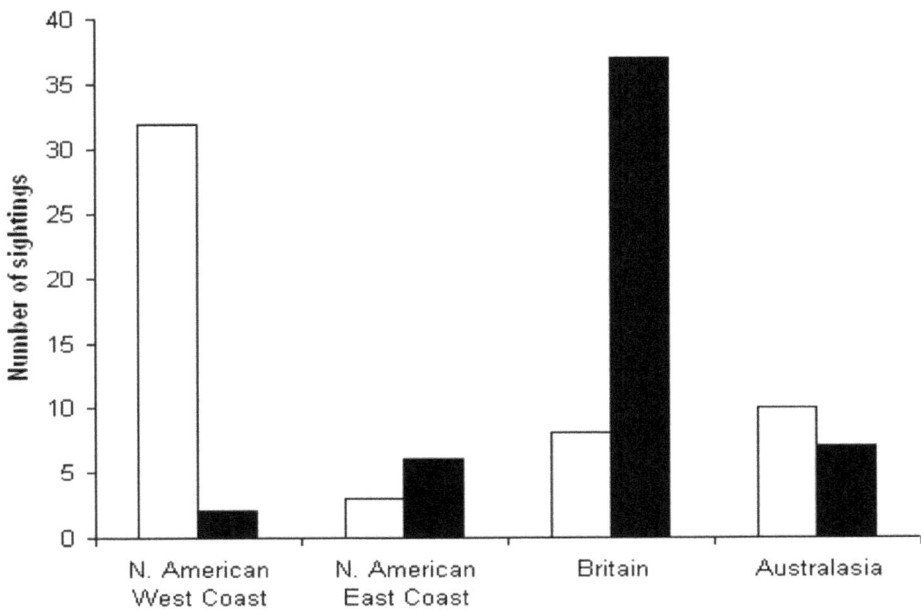

Figure 13. A graph showing the relative cumulative numbers of sightings of <u>Halshippus</u> (white) and <u>Megalotaria</u> (black) for different regions. Note that for each region the data show a clear dominance of one species over the other and that in two cases the data show considerable differences in sighting numbers indicating the possibility of strong competition in those regions (Data from Heuvelmans).

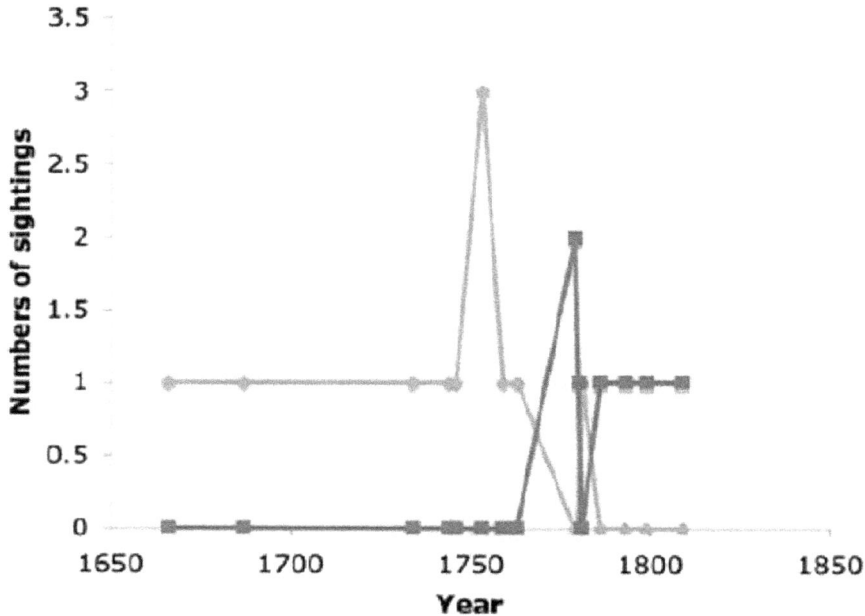

Figure 14. A graph showing the frequencies of both super-otter (circles) and <u>Megalotaria</u> (squares) sightings over a 43 year time period. Note how longneck sightings increase as super-otter sightings decrease. By the mid nineteenth century super-otters are no longer being seen, however <u>Megalotaria</u> sightings become increasingly more common from this point on (Data from Heuvelmans).

in particular), indicating that super-otters and longnecks were in competition for similar niches. It is interesting to speculate that this would seem to indicate that the pinnipeds might not have been the only ones to try and occupy elasmosaur-like niches. It is thought that the super-otter may have been driven into extinction through competitive exclusion by the rise of the longnecks, as sightings seem to have ceased after the middle of the nineteenth century.

Super-otters would not have been the only source of theoretical interspecies competition. There is a list of species, which could theoretically compete over resources and habitat with longnecks and even predate them. Cetaceans, elasmobranches (sharks), giant and colossal squids (In 1923, a *Halshippus* was supposedly spotted battling a giant squid in Nouméa Bay), [10] other pinnipeds and even humans are all potential competitors and predators.

3.4 Longnecks as *another* pinniped family?

Heuvelmans believed *Megalotaria* to be an otariid, primarily by virtue of its neck length, and its capacity for land based locomotion. Otariids have fairly long necks and slender heads, which are an adaptation to hunting. If the 'horns' of longnecks are in fact modified pinnae, then as has been suggested, this would seem to be strong evidence for their being otariids. However, Cornes has suggested that cases can be made for longnecks being in the other families as well.

He suggests that the move away from a partly land-based existence to an almost exclusively aquatic one in longnecks may be a phocid trait, as phocids are considered to be the most advanced of the pinnipeds in this respect. The presence of pinnae would seem to rule this out, as these are not typically a phocid trait.

However short weaning and maturation times *are*, which would be necessary if longnecks give birth at sea. An argument could also be made for longnecks (in particular *Halshippus*) being the evolved descendents of the swan necked seal, which is the closest thing to a true longnecked pinniped yet discovered. Cornes has suggested that a case could also be made for them being odobenids, by pointing out the uniqueness of one particular adaptation; namely their long tusks. Perhaps if this family produced one unique adaptation in the form of long tusks, then it could have produced another in the form of a long neck. In addition, odobenids combine both phocid and otariid traits, including a heightened ability for land-based locomotion.

The best case, however, seems to have been made for longnecks being in the otariid family, as evidenced by the presence of probable pinnae, elongated necks, longer fur, and the ability for terrestrial locomotion (in *Megalotaria* only). An exact taxonomic description awaits the recovery of a carcass or a live specimen, if one is ever to be found. Also, at this stage, in the absence of a specimen, it would not be wise to rule out the possibility of longnecks being part of a unique pinniped family (or even two unique families).

---

10 Although, it has been suggested recently by Marine Biologist Charles Paxton, and colleagues, that certain sightings of this nature may actually have involved genital displays in whales rather than anything new to Zoology.

3.5 Cryptic lifestyles

A reasonable question to ask is, if these large pinnipeds exist, then why are carcasses not routinely recovered, or detailed recordings made, that would allow zoologists to definitively identify them? Obviously this is a very pressing question, and the final word on the existence of these creatures will only be made with the discovery of an unambiguous carcass, or living specimen.

It is possible to speculate however as to what may make them so elusive. Firstly these pinnipeds may spend a considerably greater percentage of their time in the ocean, or other large bodies of water, than other pinnipeds. This is not an unreasonable assumption to make. Given their large size, an aquatic environment would provide their frames with much more in the way of mechanical support than a land-based one. This, coupled with the ability to give birth and breed within a body of water, could mean that these pinnipeds are rarely ever found on land, so are therefore correspondingly less likely to be found in general.

There exists a highly elusive, and rarely seen, species of pinniped known as the Mediterranean monk seal (*Monachus monachus*) whose lifestyle may provide clues as to the elusive nature of longnecks. This ultra-endangered pinniped takes refuge from predators in inaccessible underwater cave systems, where it gives birth, and rears its young. The Mediterranean monk seal spends most of its time hiding away in such caves, and is therefore rarely if ever seen. Other pinnipeds are also known to make use of caves for breeding and gathering purposes. Perhaps the shy longnecks *also* sometimes make use of caves and underground systems?

Many pinniped species, even ones that are fairly visible in their habitats, and have been well studied have cryptic elements to their lives, which seem to defy attempts to investigate. This goes to illustrate just how precious little is *really* known about even the most highly visible creatures on the planet.

3.6 Conclusions

The longnecks would seem to exhibit many traits that can find analogies within the known pinnipeds.

This seems to provide a strong case for them *being* pinnipeds first and foremost, and secondly it allows for the longnecks - in the absence of specimens - to be placed within a particular family of pinnipeds, namely the otarridae, with a fair degree of certainty.

Heuvelmans found that it wasn't easy to provide identifications for many of the sea serpents that he attempted to classify, and some of his attempts to explain away this, that, or the other sea serpent variety as a reptile, or as a possible archaeocete seem contrived at times. It is fortunate, therefore, that the longnecks can be so unambiguously allied with the pinnipeds. The case for their existence is potentially the stronger for it.

4. SIGHTINGS

4.1 Overview

So far, only the theory behind longnecks has been considered as opposed to hard observational evidence. The purpose of this section will be to present a small sampling of the more convincing evidence for the existence of longnecks, in as balanced a manner as possible. This evidence will take the form of sightings, their corresponding reports, drawings, and photographs. These data will not be presented uncritically; a sceptical approach will be used in evaluating them.

4.2 Close encounters

In this section, eight accounts, coupled with drawings and photographs where appropriate, have been selected to represent a broad spectrum of the available data on the existence of longnecks, ranging from accounts that are ambiguous and possibly fabricated, to ones that are probably genuine.

This sample is not representative of the convincing data as a whole. Heuvelmans for example collected and collated dozens of sightings data pertaining to longnecks of both varieties between the mid-seventeenth and twentieth centuries which he regarded as unquestionably genuine, however this spread of data is meant to be illustrative of both the 'spectrum of ambiguity' and the attendant issues in dealing with cryptid sightings in general.

4.2.1 Encounter number one

Year: 1555
Location: Norwegian coast
Number of witnesses: Multiple, unknown.

In his 1555 work, *History of the Northern Peoples*, the Swedish ecclesiastic Olaus Magnus [†] provided a thorough although undoubtedly exaggerated account of what is now referred to as *Halshippus olai-magni*.

"Those who sail up along the coast of Norway to trade or to fish, all tell the remarkable story of how a serpent of fearsome size, 200 feet long and 20 feet wide, resides in rifts and caves outside Bergen. On bright summer nights this serpent leaves the caves to eat calves, lambs

---

† **Olaus Magnus, Olaus Magni** or **Olaus Magni Gothus** (*Magnus*, Latin for the Swedish *Stor*, great, is a personally taken Latin family name, and not a personal epithet. His real name was Månsson; son of Måns, and was reportedly born in October 1490 in Östergötland, and died on August 1, 1557, was a Swedish ecclesiastic and writer, who did pioneering work for the interest of Nordic people. Born in Skeninge in October 1490. Like his elder brother, Sweden's last catholic archbishop Johannes Magnus, he obtained several ecclesiastical preferments. Among them a canonry at Uppsala and Linköping, and the archdeaconry of Strängnäs. **From Wikipedia, the free encyclopedia**

and pigs, or it fares out to the sea and feeds on sea nettles, crabs and similar marine animals. It has ell-long hair hanging from its neck, sharp black scales and flaming red eyes. It attacks vessels, grabs and swallows people, as it lifts itself up like a column from the water."

Figure 15. *The great 'Sea Orm', as illustrated by Olaus Magnus in his History of the Northern Peoples (1555).*

This description obviously exaggerates the proportions of *Halshippus*, as it suggests a length of 200 feet and a body diameter of 20 feet, which would make what is being described a good 100 feet longer than the longest reliable estimate of *Halshippus'* length. Also it is suggested that it is capable of land-based locomotion in order to find prey; again this is inconsistent with what can be inferred about *Halshippus*. However, certain key details leave little doubt as to the identity of the 'Sea Orm'. For example, it is described as having '*ell-long hair hanging from its neck*' and '*flaming red eyes*'. These descriptions correlate very strongly with similar characteristics observed in other *Halshippus* sightings.

Another interesting point to note, is the aggression that it allegedly exhibits. It was mentioned in section three that a *Halshippus* was spotted battling a giant squid in 1923. Sanderson has suggested that these creatures are in fact rather aggressive (when they grow over a certain size perhaps), based on the fact that natives of the Arafura Sea, where many *Halshippus* sightings have been made, describe them in their own language as the 'enemies of giant squids'. They apparently attempt to mutually prey upon one another in that region.

Of course, it must be noted that even the most competent naturalists of the sixteenth century had little accurate knowledge concerning most marine life. The seas and oceans were still largely unknown areas of danger, and the fantastic creature described by Olaus Magnus could simply be a collective distillation of the fears of every sailor and traveller of that time.

4.2.2 Encounter number two

Year: 1879
Location: The Gulf of Aden.
Number of witnesses: Three.

Major H. W. J. Senior, of the Bengal Staff Corps, Miss Greenfield and Dr. C. Hall, were witness to a long-necked creature, whilst on board the *City of Baltimore* in the Gulf of Aden. The following is the Major's report, which was countersigned by two witnesses. The full account can be found in Heuvelmans (p. 280).

"So rapid were its movements that when it approached the ship's wake, I seized a telescope, but could not catch a view as it darted rapidly out of the field of the glass before I could see it. I was thus prevented from ascertaining whether it had scales or not but the best view of the monster obtainable when it was about three cables' length, that is about 500 yards' distant, seemed to show that it was without scales. I cannot, however, speak with certainty. The head and neck, about two feet in diameter, rose out of the water to a height of about twenty or thirty feet, and the monster opened its jaws wide as it rose, and closed them again as it lowered its head and darted forward for a dive, reappearing almost immediately some hundred yards ahead. The body was not visible at all, and must have been some depth under water, as the disturbance on the surface was too sleight to attract notice, although occasionally a splash was seen at some distance behind the head. The shape of the head was not unlike pictures of the dragon I have often seen, with a bull-dog appearance of the forehead and eyebrow."

Figure 16. *The long necked sea serpent as viewed from the deck of the* City of Baltimore. *From Oudemans, (1892).*

The wishful alikening of what was seen to a dragon aside; elements of this report correspond to other longneck sightings. The bulldog like face for example can be compared with other witnesses to longnecks who have described the heads as being 'dog-like'; this also indicates a degree of mammal-ness in what was being seen. It is also worth noting that the Major, despite his search, could not discern reptilian characteristics in the creature (although as he himself admits the circumstances under which the creature was being viewed were hardly ideal for scrutiny).

The opening and closing of the jaws is an interesting detail, as - perhaps - the creature was posturing in response to a perceived threat from the *City of Baltimore*? Another interesting detail is the lack of 'horns'; an otherwise common component of longneck sightings. Perhaps they were flattened against the creature's neck, or maybe they were absent altogether?
It is certain that *something* was seen, but the question is what? A good case can be built for this being a longneck, as was just demonstrated. However, a case could also be made for the object being a tree trunk bobbing in the water, so as to give the appearance of a long-necked creature diving and rising. The remainder of the details could be the products of collective false memory on the part of the witnesses. The *City of Baltimore* sighting is curious, and in many ways convincing, but like all sightings where there is an absence of stronger evidence, it has to be regarded as fairly subjective.

4.2.3 Encounter number three

Year: 1913.
Location: The Grand Banks, Newfoundland.
Number of witnesses: One.

G. Batchelor, an officer serving onboard *The Corinthian*, made the following report, which was published in Heuvelmans' *In the Wake of Sea-Serpents*. The report is included here exactly as it appears in Heuvelmans' book (p. 392-394).

"As the 'Corinthian' was ploughing her way westward, I was officer of the watch 'on duty at the time'. At 4.30 a.m. in the cold grey dawn of August 30th, 1913, on The Grand Banks of Newfoundland, the look out man had just gone off and the third officer had left the bridge to see if all was well around the decks, while casting my eyes around the horizon I picked up an object about a mile off right ahead. The best conjecture I could make as to its nature was that it was a fishing boat lying end on to us. In the dense and extensive fogs, which sweep over the fishing banks sailors frequently become separated from their schooners and many starve for days before being picked up. I had just such an incident in mind as I watched the object ahead. When it suddenly disappeared beneath the surface, being still unenlightened I thought of tragedy.

Suddenly however, after I had meditated upon serious things something surprising showed itself about two hundred feet away from the ship. First appeared a great head, long fin like ears and great blue eyes. The eyes were mild and liquid, with no indication of ferocity. Following sad eyes came a neck, it was a regular neck alright, all of twenty feet in length, which *greatly resembled a Giraffe. The monster took its time in emerging so long that I wondered what the end would be. The neck...seemed to be set on a ball bearing, so supple was it and so rhythmically did it sway while the large liquid blue eyes took in the ship with a surprised, injured and fearful stare. The creature was well fixed for side arms. Three horny fins surmounted its bony head, probably for defence and attack or for ripping things up. The body was about the same size as the neck very much like a monster seal or sealion with short water smoothed fur. The tail was split into two fins. The colour scheme was good, although some might think it giddy; light brownish yellow tastefully spattered with spots of a darker hue. For a minute the creature inspected the 'Corinthian' with its roving gaze, and then disappeared, showing its afterworks as it dived. Its whole attitude while in sight was that of one `moving about in worlds unrealised`. It seemed to be trying to comprehend a curiosity, which it had good reason to believe, might be a new danger. I almost felt a tenderness for it, and never have I experienced such a minute in my life. Down in my room I had a camera and a rifle. Yet I was the only one on the bridge besides the quartermaster at the wheel. I don't mind confessing that I wavered between my duty and desire for some sort of shot. Finally I stayed, but I don't know whether I should take full credit for that or not because I hated to lose sight of the thing. As it watched me it churned the water into foam and spray with its huge front fins. As it went out of sight it emitted a piercing wail like that of a baby. Its voice was altogether out of proportion to its size.*

Figure 17. Batchelor's enigmatic illustration of what he saw. From Heuvelmans (1968).

This report, despite the imaginative prose with which it was written, is quite interesting as it reveals details such as a long neck, the presence of ear and horn-like structures (pinnae possibly), forelimbs, fur, and a tail 'split' into two hind-limb flippers. However the extraordinary degree to which Batchelor goes in anthropomorphizing his sighting ('*The eyes were mild and liquid*', '*sad eyes*', '*injured and fearful stare*' etc.) and the highly dubious zoological conclusions to which he jumps [11], suggests that he intended to sensationalize the account, which unfortunately goes someway to discrediting it as evidence. Even the illustration that he made after the fact, seems to show dragon-like embellishments, and a peculiar 'beard' which apparently resembled his own (See figure seventeen).

The most important thing that has to be deduced from this is what exactly it was that he was sensationalizing? Was it something made up, something mundane or was it a genuine longneck?

The only thing that suggests that Batchelor may have seen a genuine longneck (*Megalotaria* in all likelihood, given the location of the sighting, and the shreds of useful information that can be teased from this account) is that the small aforementioned interesting anatomical details he described, seem to correspond to other accounts that can be considered less ambiguous and more reliable.

Heuvelmans suggested that this account was probably genuine, partly on the strength of third party testimonials to Batchelor's character.

4.2.4 Encounter number four

Year: 1919.
Location: Orkneys, Hoy (UK).
Number of witnesses: Multiple, unknown.

J. Mackintosh Bell, a lawyer and writer was helping some friends on a local fishing boat, the crew of which had reported seeing a long-necked sea serpent in the area on several occasions. The said same longneck made an appearance whilst Bell was present. This resulted in the following report, which was communicated to Rupert T. Gould, and later appeared in Heuvelmans' book. It is repeated here exactly as it appears (p. 402-404).

"I looked and sure enough about 25-30 yards from the boat a long neck as thick as an Elephants fore leg, all rough looking like an Elephants hide, was sticking up. On top of this was the head which was much smaller in proportion, but of the same colour. The head was like that of a dogs, coming sharp to the nose. The eye was black and small, and the whiskers were black. The neck, I should say, stuck 5-6ft., possibly more out of the water. The animal was very

---

11   Batchelor suggests in an un-reproduced portion of his report, that the monster was a plesiosaurus – despite attributing to it mammalian characteristics. He also suggests that the appearance of the creature had something to do with the presence of the wreck of the *Titanic*, which had gone down in the vicinity of *The Corinthian* one year previously

shy, and kept pushing its head up then pulling it down but never quite going out of sight. The body I could not see. Then it disappeared and I said `If it comes again I'll take a snapshot of it'. Sure enough it did come and I took as I thought a snap of it, but on looking at the camera shutter, I found it had closed owing to its being swollen, so I did not get a photo. I then said 'I'll shoot it', but the skipper would not hear of it in case I wounded it, and it might attack us. It disappeared and as was its custom swam close along side the boat about 10ft. down. We all saw it plainly, my friends remarking that they had seen it swimming just the same way after it had shown itself on the surface. My friends told me they had seen it in the same place the year before. It was a common occurrence so they said. That year was the last of several years in which they saw it annually. It did not show itself for two or three years, and then it was only seen once. As to its body, it was, as seen below the water, dark brown, getting slightly lighter as it got to the outer edge, then at the outer edge appeared to be almost grey. It had two paddles or fins at its side and two at its stern. My friends thought it would weigh 2 or 3 tons, some thinking 4-6. Not only my friends, but others lobster fishing, got many chances of seeing it.

Dimensions; Neck, so far as seen, say 6-7ft. Body, never seen when neck straight up, but just covered in the water. You could detect the paddles causing the water to ripple. When underwater, swimming, the body, I think, to the end of the tail flappers would be about 12 feet long, and if the neck was stretched to 8 ft, the neck and body 18-20ft. long. The skipper of the boat remarked that sometimes the top of the head, when seen from a boat vertically, was a bright red. Neck thickness say 1 foot diameter: Head very like a black retriever say 6" long by 4" broad. Whiskers black and short. Circumference of body say 10-11 ft., but this I am not sure of, as I never saw all round it, but it would be 4-5ft. across the back."

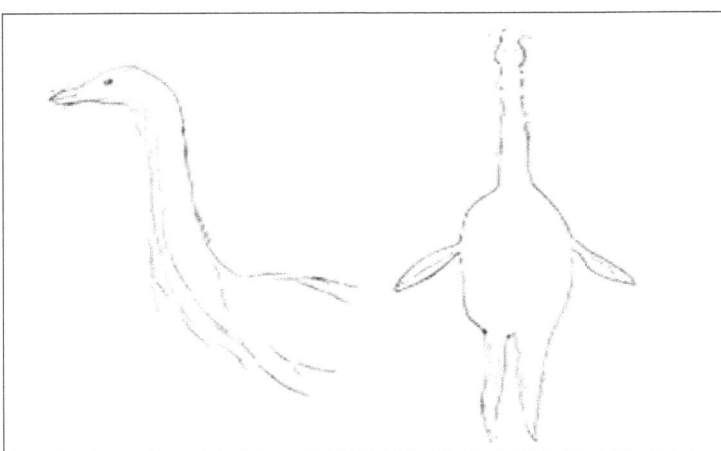

Figure 18. *Macintosh Bell's sketches that accompanied his description. From Heuvelmans (1968).*

This report is especially interesting for a variety of reasons. Firstly, there is the level of objective detail that is being reported. Due to the proximity of the witness from what he is witnessing, much can evidently be seen, including details of the head and neck, the body (including the presence of fore and hind flippers) and intriguing colour gradations. It is a shame that Bell's camera was not functioning at the time, as a photograph would have been of infinitely more value than the scratchy sketch that accompanied the report. Then again the non-functioning camera does seem to be a very convenient excuse.

Heuvelmans regards Bell's character to be reliable; so, in part on that basis, the report is considered to be probably genuine. If what was seen was a long necked pinniped, then this report represents one of the most clear, detailed, and convincing accounts obtained to date.

4.2.5 Encounter number five

Year: 1934.
Location: Yorkshire, Filey (UK).
Number of witnesses: One.

Wilkinson Herbert, a coastguard, was walking along the beach one night when he encountered something untoward. The following report was published in the *Daily Telegraph* (01/03/1934 edition). It also appears in Heuvelmans' book (p. 456).

"Mr. Wilkinson Herbert, a Filey coast guard, says he saw the thing on shore last night; a dark moonless night. He was walking along Filey Brig, a long low spur of rocks running into the sea, when; 'Suddenly I heard a growling like a dozen dogs ahead, walking nearer I switched on my torch and was confronted by a huge neck, six yards in front of me, rearing up 8ft. high! The head was a startling sight-huge eyes like saucers, glaring at me, the creatures mouth was a foot wide and neck would be a yard around. The monster appeared as startled as I was shining my torch along the ground I saw a body about 30ft. long. I thought this was no place for me and from a distance I threw stones at the creature. It moved away growling fiercely and I saw the huge black body had two humps on it and four short legs with huge flappers on them. I could not see any tail. It moved quickly, rolling from side to side, and went into the sea. From the cliff top I looked down and saw two eyes like torchlights shining out to sea 300 yards away. It was a most gruesome and thrilling experience. I have seen big animals abroad, but nothing like this.'"

This report is intriguing in its details, like the others, but also like the others there are elements to the account which mean that it should also be treated sceptically. As has been mentioned, the details paint a convincing picture that a sea serpent of the long-necked pinniped variety was indeed seen. The body lengths as reported by Herbert would seem to put the creature in the correct size bracket for an adult - or near-adult - longneck (of the *Megalotaria* variety). One detail stands out in particular; namely the absence of a tail. Of all the potential dissimilarities between elasmosaurs and longnecks, the absence of a tail in the case of longnecks is one of the most prominent. The Batchelor account and the crudely drawn image provided by Mackintosh Bell as part of his report - see figure eighteen - also indicate that there were no tails on the creatures that they saw either - making this a significant point of correlation.

However, there are elements of this report that need scrutinizing. For example, the report was made at night, where mundane objects are more likely to be mistaken and/or exaggerated. Giving Herbert the benefit of the doubt, could he have been the victim of a hoax? Could he have perhaps mistaken a smaller animal for a larger one? Or could he have actually encountered a longneck?

4.2.6 Encounter number six

Year: 1956.
Location: The Discovery Light, British Columbia.
Number of witnesses: Two.

David Miller and Alfred Webb, two commercial fishermen, were fishing off the Discovery Light, when they encountered something unusual. Miller made the following report, which the Oceanographer Paul LeBlond first recounted in an article he wrote for the *Montana Magazine* in 1993 (p. 44-51).

"I observed this strange creature surface roughly 80 ft. from our port beam. It started to move rapidly away from us so we speeded the engine up and gave chase. We got within 30ft. when it suddenly submerged, not in the method that seals and sealions do, but as though something had pulled it under. A few minutes later we arrived at the place of submergence and there was turbulence suggesting a 30ft Sea Whale. Its speed underwater was also astounding as it surfaced a few minutes later over a hundred yards away. It stayed up while we took off after it again but this time it never let us get close again. The first encounter was so close that both of us remarked about its large red eyes and short ears visible at that range."

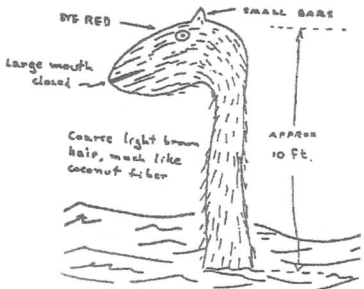

Figure 19. *Miller's sketch of what he saw. Note the 'small bars', which are in the wrong place for them to be breathing tubules, but in approximately the right place for them to be pinnae. From LeBlond and Bousfield (1995).*

This report is attributed to a cryptid known as 'Caddy', or the Cadborosaurus, which is thought to reside in, and around, the coastal waters of British Columbia.

Accounts of 'Caddy' place it in the lower size range for a *Halshippus*. In addition it is described as being very fast, sometimes having a main of fur, a 'horse' or 'camel' like head, red eyes (compare with Olaus Magnus' account) and possibly pinnae (all of which are evident from the Miller report).

Sightings of 'Caddy' go back a long time, and references to it figure in Native American traditions from the area. More recently, however, within the last 200 years, approximately 300 sightings have been made, and there are eleven reports of reputed specimens of this creature being stranded or captured (and on one occasion even being removed from the stomach contents of a whale). But unfortunately, save a handful of ambiguous photographs, the physical evidence seems to have vanished every time.

Figure 20. *A 'Caddy' carcass recovered from the stomach contents of a whale photographed in October 1937. Shortly after it was photographed the carcass vanished.*

The most prominent work on this particular cryptid is the book *Cadborosaurus: Survivor from the deep,* which was co-authored by the Oceanographer Paul H. LeBlond and the Marine Biologist Edward L. Bousfield. These Scientists made news in 1995 by publishing a paper in *Amphipacifica: Journal of Systematic Biology*, which suggested that there was sufficient 'in-hand' evidence to recommend the formal taxonomic recognition of this species. LeBlond and Bousfield have suggested the formal scientific name *Cadborosaurus willsi* for this cryptid, whilst also suggesting that it is in fact a variety of reptile.

The problem with the reptile theory of 'Caddy' is that nearly everything about this creature indicates strongly that it is not a reptile. Firstly it has mammalian characteristics; for example mains of hair have been reported, in addition to external 'horns' (pinnae). These creatures have snake-like bodies which seem to be mostly mobile through the vertical plane and, upon examination of reports of 'in-hand' specimens and photographs, it has been revealed that they possess flipper-like fore-limbs and fused hind-flippers with no tail (a pinniped trait). Also the waters of British Columbia are cold, which - as has been mentioned - would seem to make such a habitat unsuitable for reptiles.

Heuvelmans believed 'Caddy' to be a long-necked pinniped of the 'Merhorse' (*Halshippus olai-magni*) variety. At the very least 'Caddy' may be a new species within the genus *Halshippus*. That it is a mammal and a pinniped in particular seems beyond question at this juncture in the absence of a physical specimen to say otherwise.

4.2.7 Encounter number seven

Year: 1976.
Location: Rosemullion Head, Falmouth Bay Cornwall.
Number of witnesses: One.

In 1976, a witness whose only given name was 'Mary F' submitted two low quality photographs of what appears to be a long necked sea serpent to the *Falmouth Packet*, a Cornish periodical which featured the images on its front cover, claiming that they were of the legendary Morgawr – a sea serpent native to Cornish waters.

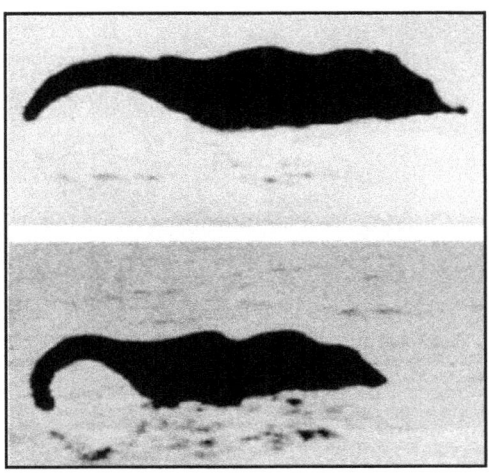

Figure 21. 'Morgawr' as photographed by 'Mary F'.

EDITOR'S NOTE: In my book *The Owlman and Others* (CFZ, 1997) I present a body of evidence which suggests that these photographs were faked by the late John Gordon of Falmouth, as part of an ongoing surrealchemical practical joke. Whilst there is no way that these pictures should be considered genuine cryptozoological evidence, they are interesting from a psychosocial point of view, and are included in the present volume for this reason. Just because these pictures are faked, however, does not mean that the other morgawr sightings along the coast of southern Cornwall should not be taken seriously. JD

The images were accompanied by the following description of the sighting from 'Mary F'.

"It looked like an elephant waving its trunk, but the trunk was a long neck with a small head at the end, like a snake's head. It had humps on its back, which moved in a funny way... the animal frightened me. I would not like to see it any closer. I do not like the way it moved when swimming."

Morgawr has quite an established pedigree as sea serpents go. The first sighting was made in 1876, when fishermen in Gerran's Bay, Cornwall allegedly accidentally captured a long necked sea serpent in their nets. In 1975 two witnesses claimed to have observed a 'humped' creature with 'stumpy horns' and bristles on its neck (classic longneck characteristics) catching a conger eel off Pendennis Point. There are many suspicious elements in the 'Mary F' story, for example 'Mary F' has never been located, and neither have the original photographic negatives.

The object that she photographed in the water seems to bear a suspicious resemblance to an elephant swimming in the water holding its trunk up. Also, some have suggested that in light of the 1991 *Strange* magazine revelations [12], Shiels may have been responsible for hoaxing the 'Mary F' sighting [13]. Even if the most significant 70s Morgawr sighting was a hoax, that still leaves other sightings made during this decade and sightings of previous and latter decades, whose authenticity has yet to be called into question. [14]

4.2.8 Encounter number eight

Year: 1977.
Location: Lake Champlain, upstate New York.
Number of witnesses: Four.

Sandra Mansi, her fiancé, and Sandra's two children from a previous marriage, were on a trip to Lake Champlain, in upstate New York. They stopped, and Mansi's children went to play in the lake. Mansi went to get a camera to photograph them, whereupon she spied a long-necked creature of some size, fifty yards out in the lake from where her children were playing. She made sure that her children were out of potential harm's way, before snapping the now famous photo of what she saw.

---

12 However, a 1991 *Strange* magazine exposé of a series of transcripts taken from some tapes that Shiels made, could be interpreted as being practically tantamount to an admission of a hoax on his part.
13 Although given what was revealed in the transcripts, it seems more likely that - if they were faked - Shiels hoaxed only his own sighting, and had nothing to do with 'Mary F's'.
14 It is intriguing that there appeared to be so much long neck activity in the Atlantic region during the 70's. In addition to the rash of Morgawr sightings made during this decade, not all of which could have been hoaxed, the famous Mansi photograph of 'Champ' was taken (which will be described fully in 4.2.8). This could be evidence for a continuation of the visual correlation suggested in figures ten and eleven, where catch sizes for Atlantic cod species increased sharply during this decade.

She described the creature as having had a long neck and head, which can clearly be seen on the photo, as well as what appears to be a body breaking the surface.

She says that the creature dived underwater shortly after she photographed it, and it did not surface again.

Figure 22. The famous image of the Lake Champlain cryptid.

The Lake Champlain cryptid, 'Champ' or *Champtanystropheus*, as it is sometimes known, is rarely ever seen, especially when compared to other lake monsters such as Ogopogo or the Loch Ness monster. However unlike these other alleged lake monsters, Champ actually has some tantalizing photographic evidence, which - as of yet - has not been demonstrated to be fraudulent. As a matter of fact it has often been said of the Mansi photograph that it is by far the strongest piece of supporting evidence for the existence of lake monsters acquired to date. What's more, this photograph appears to show something that resembles a longneck. The head is held high out of the water vertically, a mammalian trait, and the neck and what appears to be body breaking the surface are brownish – a common characteristic of longneck sightings. If this is a longneck, then it could be expelling air before diving, or maybe even looking for others of its kind (which may be the case, if it has come to the lake to breed).

An analysis by LeBlond indicated that the object in the photo may have been between sixteen and fifty-six feet long, putting it in the right size range for an adult longneck.

Another interesting factor to note about 'Champ' is that it is described as 'horned' in the traditions of the Native American tribes, who have lived in the area for millennia. However no horns are visible in the Mansi photo (perhaps they are flattened against the creatures head?) However, the sceptics have some salient words of caution regarding this particular photograph. The exact location at which the photograph was taken remains unknown, also there is very little in the photograph that would allow for meaningful size comparisons with other objects to be made.

Benjamin Radford, for the *Skeptical Inquirer,* conducted a recent analysis of this photograph which indicated that some of the preliminary size estimates for the creature may be wrong, and that whatever is in the photo is probably smaller than has been reported. He concludes however, that Mansi counts as a credible witness to *something* in the water, although what exactly that may be, is obviously open to debate. He believes that the object was a tree stump.

The possibility of a hoax (perpetrated by someone other than Mansi and her family) is also something that still needs to be considered, as is the possibility of misidentification. However as it stands, the Mansi photograph may be one of the strongest single pieces of evidence gathered so far, not just for the existence of lake monsters, but for the existence of long necked pinnipeds in general.

4.3 Summary

Sightings, photographs, and carcasses, make for tantalizing evidence, but there are limits as to their value as data. Sightings data, for example, are mostly subjective, as they are always reported after the fact, and are therefore subject to any number of distorting processes, including omission, commission, exaggeration etc.

In this respect, cryptozoology is a partly forensic science, and as in forensic reconstructions of crime scenes, the only ways of increasing the potential value of witness data are through the use of character testimonials (making sure that the individuals involved are not habitual liars for example) and in the comparison of different reports of both single and unrelated sightings, to see if there are any noteworthy correlations.

Of course this process is still error prone, as - for example - not being a pathological liar is not going to prevent an otherwise reliable witness from remembering a particular detail incorrectly. Also different sightings with similar details (which may indicate that what is being seen in one location is also being seen in others) may be due to the collective expectations of how a monster *should* look formed by the media and popular perception on the part of unconnected groups of witnesses.

Photographs make for more tantalizing evidence than simple sightings reports, and carcasses even more tantalizing still. However these evidences *still* have their limitations, as in the case of photographs, details of location are all important. In the absence of these, even a photo which passes many analytical tests, can have surprisingly little value as evidence [15].

Carcasses, like those that were found and photographed of 'Caddy', have the potential to function as irrefutable evidence, or at least this would have been the case, if the carcasses hadn't gone missing, or if people would think to take samples rather than simply photographs as this would provide conclusive data on the subject.

---

[15] As can be seen with the issue of sighting location and the Mansi 'Champ' photo.

Despite the subjective limitations of these evidences however, it would be unwise in the extreme to simply dismiss them all out of hand on that basis, as do many sceptics. Witness sightings reports, carcasses, and photographs, all posses a signal to noise ratio of useful data. The subjectivity of these data represent 'noise', and through cross-comparison, character testimonials, detailed analysis, and correlation with extrinsic factors (as with the plausibility method) any 'signal' of usefulness can be effectively teased from the data. On this basis they can still function as evidence, whose value depends on the signal to noise ratio, however they must *always* be discussed in the context of their limitations as data. Not to do so is scientifically dishonest.

5. WHAT DOES THE FUTURE HOLD IN STORE?

5.1 Overview

Having reviewed the opposing identity theory arguments, and having looked at the evolutionary biology, theoretical ecology, and the sightings data, a fairly good case can be made for the existence of long-necked pinnipeds exhibiting a large degree of niche breadth (being able to make use of multiple habitats for different purposes). Their current resistance to being found, and formally identified, may be due to a combination of factors, including - but not limited to - their comparative rarity to other species, their shyness, their highly aquatic-centred life style, and their possible use of out of the way or underwater cave systems, which would effectively render them invisible to detection.

The *least* ambiguous piece of supporting evidence for their existence is currently the Mansi 'Champ' photograph, which - as was mentioned in the previous section - has yet to be demonstrated conclusively to be either a hoax or misidentification. Although 'in-hand' specimens of 'Caddy' (a probable *Halshippus*) have allowed a sufficiently strong case to be made for its tacit recognition as a new species in the peer reviewed scientific literature, LeBlond and Bousfield, the scientists behind this tremendous accomplishment seem to have misidentified 'Caddy' as a reptile, and in doing so have (in the opinion of this author) passed up the opportunity to gain the first formal scientific recognition for a long-necked pinniped species.

All it will take is a single unambiguous carcass, a single piece of high quality film, or the capture of a live specimen for these creatures to gain formal scientific acceptance, but until then, as has been established, there are still grounds for speculation.

5.2 Conservation issues

Obviously, creatures as infrequently seen as longnecks, are going to be fairly rare, so they are thusly going to require some kind of conservation effort in order to prevent extinction (assuming it is not already too late). It appears that unlike formal acceptance by zoologists, not much is required for a potential species to gain protection. After 'Champ' was photographed, both the states of Vermont and New York added it to their lists of protected species - so it would seem that longnecks at least enjoy some measure of protection in those states.

LeBlond and Bousfield recommended in their *Amphipacifica* article that 'Caddy' be placed on a protected species list, due to dwindling numbers of sightings over the years. It is unknown if the Canadian government has made any effort to do so however.

The problem with the current conservation effort for longnecks as it stands, is that regional protection would appear to be inadequate given what can be inferred about them. Longnecks seem to be cosmopolitan in their distribution, and make use of lakes possibly to breed, or to reduce competition pressure. The protection afforded longnecks in Lake Champlain doesn't protect them at sea, and longnecks are likely to spend much more time there than anywhere else.

For global protection and inclusion in CITES, more concrete evidence must be found to support their existence. But despite this it may be interesting to speculate on the form that potential obstacles to conservation may take.

5.2.1 Over-fishing

It is estimated that currently the world's oceans are being over-fished, which means that the rate at which fish are being removed from the oceans is exceeding the rate at which stocks are replenishing. Many - once abundant - species of fish are now very rare due to this practice; a good example of this being the blue fin tuna, which was once abundant and cheap, but is now expensive and rare. In section three a graph showing the change in longneck sightings over time, was compared to one showing Atlantic cod abundance data, and it appeared that there was a loose visual correlation between the two. If this correlation is due to there being any actual relationship between the two sets of data, then the suggested decline in longneck sightings over recent decades, post-Heuvelmans, would seem to indicate that their numbers are being influenced in a detrimental fashion by over-fishing.

Nets and fishing lines are another potential problem for longnecks, and sea creatures in general. Large numbers of species with no commercial value are caught in nets and lines, and discarded. There is a precedent for longnecks being the victims of nets. A reputed accidental netting of a longneck off of the coast of Cornwall was mentioned briefly in the last section. There have also been other reputed incidents of longnecks getting entangled in fishing nets. It would seem however that nets and lines do not pose as big of a problem for longnecks as for dolphins, for example, as if they were caught with the same frequency, then there should be some material evidence of their existence by now. One possibility is that if longnecks are people averse, and as they appear to prefer coastal habitats (although *Halshippus* seems to be adapted for a more semi-abyssal lifestyle), then they are simply less likely to get caught by fishing vessels that tend to patrol the open oceans. Another possibility, is that if longnecks are caught at sea, the incidents of this go largely unreported, but this seems quite unlikely.

5.2.2 Increasing ocean traffic

As was mentioned in section three, it has been suggested that the decline in longneck sightings post-Heuvelmans, can be accounted for in part, by the fact that the volume of ocean traffic has been increasing in recent decades. Longnecks, being probably very shy would naturally seek quieter regions, however if suitable habitat is being increasingly disrupted, then it is likely that their numbers would suffer under ecological stress.

5.2.3 Pollution

Another factor that could pose ecological problems to populations of longnecks is pollution. Assorted chemical effluvia from land-based sources, and the end products of internal combustion, are being deposited into the oceans at a fairly high rate. If longnecks, or their primary food sources, are sensitive to pollutants, then this could conceivably have a long-term effect on their chances for survival.

5.2.4 Climate change

Increases in the global mean temperature (whether mostly due to anthropogenic, factors or mostly due to natural ones, or - as seems more likely - due to some currently unknown combination of the two) have the potential to detrimentally effect fragile longneck populations.
An increasing oceanic temperature may result in marine ecologies changing significantly over time, carnivorous secondary and tertiary consumers like pinnipeds are going to be amongst the most susceptible to these changes, as they are the least ecologically efficient, and are most reliant upon complex marine food webs remaining intact at their lowest trophic levels.

5.3 A conservation plan

Given that potential obstacles to the long - term survival of longnecks have been identified, it would seen logical to suggest that a reduction in things like over-fishing and pollution levels would be a suitable conservation plan for these creatures. This is partly true, and the hope is that current conservation efforts aimed at having an effect on these factors, would be of collateral benefit to longnecks. In order to be optimal however, a conservation plan aimed at longnecks in particular, will have to take into account their particular idiosyncrasies. For example, if longnecks *do* move between ocean and freshwater lakes, then it would be a good idea to ensure that their potential points of access to - and exits from - these bodies of water remain as unimpeded as possible. This would be of particular significance if longnecks use freshwater lakes to breed.

5.4 Chances of discovery

The bottom line is that for longnecks, the chances of future discovery and formal identification would seem to be very much dependent on the success of extant marine conservation programs; which to have their maximum beneficial effect on longneck populations, would need to be tailored to them, which in turn would require formal recognition and CITES listing - making this somewhat of a `Catch - 22` situation.

Longnecks are extremely rare, and always have been. Heuvelmans' prediction that long necked pinnipeds were on the rise in general has been shown to be more than likely incorrect. The reality is that these animals probably comprise small, ecologically stressed, and rapidly dwindling populations.

Charles Paxton, who along with colleagues, proposed the theory that some sea serpent sightings can be accounted for by cetaceans in a state of arousal (see footnote ten), suggested that based on a cumulative description curve for large (two metres or more along a major axis) open water animals, there were approximately forty seven species yet to be described (as of 1998), as his curve had not reached an asymptote. Based on this, Paxton predicts that new, large marine animals are discovered and described on average once every 5.3 years. He also predicts that the majority but *not* all of the species that are awaiting discovery are cetaceans. This is interesting as here, for the first time, is a predictive indicator of what is left to discover, and it appears that longnecks may simply be waiting their turn.

Should the ultimate discovery be forthcoming, then a new and exciting chapter in the history of zoology can be written, as it was with the coelacanth and megamouth shark. But should a major discovery *not* be made soon, then there is every chance that this potential species could go extinct in the absence of a specifically tailored conservation plan. One of the main purposes for which this author would like this particular chapter to be employed, is in making the future conservationists life easier by speculating on the ecology of these creatures, and allowing for such pre-emptive conservation strategies to be devised.

CHAPTER 2.
Heuvelmans' archaeocetes

The many-finned sea serpent
The super-otter
The many-humped sea serpent

The many finned sea serpent: Time to reclassify?

1. INTRODUCTION

1.1 Overview

In this section, after reviewing contending identity theories, a case will be made for the many finned sea serpent, also colloquially known as the *con rit*, being a new member of the subphylum Myriapoda in the class Arthropleuridea. It is proposed that the *con rit* may therefore belong to a novel parataxa genus; *Mariascolpenda*, and that the primary case for its existence is made by the specimen-based descriptions provided by Mr. Van Tran Con in Hongay, Vietnam, 1883 and related in detail by Dr. A. Krempf in 1921, which constitutes both strong in-hand evidence and a type specimen.

1.2 A centipede by any other name

Despite the fact that it has been given no fewer than three common names which include the word 'centipede' - 'cetacean centipede', '*con rit*' (Vietnamese for centipede, or millipede) and even 'great sea centipede' (which was first used to describe it in the third century CE by the Roman teacher and author Claudius Aelianus); there are very few researchers who have suggested a Myriapod identity for the many finned sea serpent, and this view is largely considered heterodox within cryptozoological circles. Interestingly, several websites have quoted Dr. Karl Shuker as having suggested such an identity, and also that he suggested that it was an unknown species of giant isopod, but this is in fact incorrect. Dr Shuker *has* suggested that the *con rit* could be an unspecified unknown crustacean, but has not committed himself further.

However, when the other proposed identity theories are scrutinized, it seems - surprisingly - that a Myriapoda identity best fits the admittedly very limited data concerning this creature's morphology, characteristics, speculative evolutionary history, and ecology.

2. IDENTITY THEORIES.

2.1 Overview

The relatively few identity theories that have been advanced regarding the many-finned sea serpent (henceforth *con rit*) can be loosely divided into two categories; cetacean and invertebrate. The *con rit* is not a well-known cryptid, so therefore has stimulated little in the way of debate amongst cryptozoological researchers regarding its identity.

2.1.1 The many-finned cetacean theories

Although Aelian did not offer an identity for his 'great sea centipede' in his *On the Nature of Animals,* his writings coupled with reports of sightings were to inspire the French medical lecturer and Naturalist Guillaume Rondelet to reclassify it as a 'cetacean centipede' in his 1554 text, *De Piscibus Marinum.*

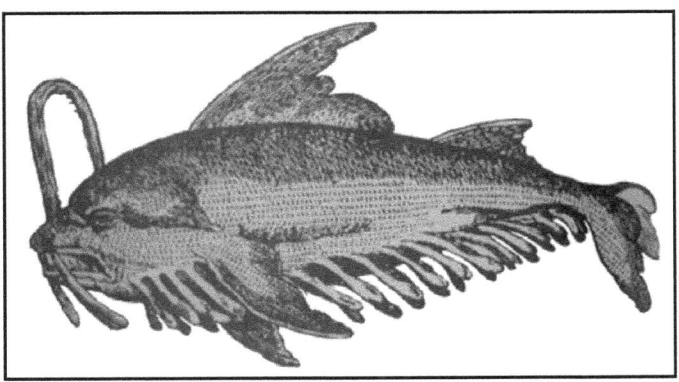

Figure 1. The cetacean centipede as imagined by Rondelet (1554).

Heuvelmans revived speculation regarding the identity of the *con rit* in *In the Wake of the Sea-Serpents,* and was the first to suggest the name 'Many finned sea serpent' for it. He postulated that the *con rit* was a 60 to 70 feet long version of the primitive cetacean archeocete basilosaurus, which - based on the speculation of a correspondent called "Howell" - may have possessed numerous lateral fins as an aid to stabilizing the hydrodynamic forces over its body during locomotion. Heuvelmans also attributed to this archeocete armour plating.

Coleman & Huyghe's speculations largely concur with those of Heuvelmans, in that they also believe the *con rit* (which they re-name the 'Great sea centipede' after Aelian) to be an Archeocete.

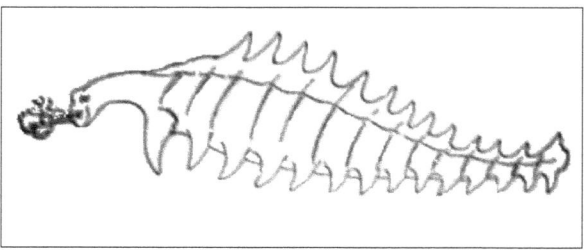

Figure 2. Heuvelmans' many-finned sea serpent (1968).

Champagne, who categorizes the *con rit* as a 'type seven' or 'segmented/multi-limbed' sea serpent, describes it as a cetacean although he is non-specific about the exact type.

Figure 3. Champagne's vision of his multi-limbed sea serpent illustrated by Cameron McCormick.

2.1.2 Invertebrate theories

The closest thing approaching a consensus in Cryptozoological circles regarding the identity of the *con rit* is that it is probably a cetacean, or at the very least a mammal of some kind. However there is a minority position within Cryptozoology which regards the *con rit* as an invertebrate; either a large member of the Isopoda (a diverse order of crustaceans) or an undiscovered aquatic species of Chilopoda (centipede).

Researcher Lance Bradshaw, has also suggested giant polychaete worms as a possible identity.

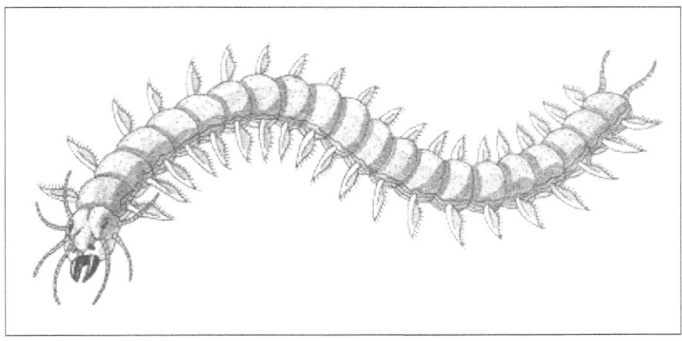

Figure 4. Bradshaw's depiction of the con rit as a polychaete worm.

3. SIGHTINGS AND CARCASSES

3.1 Descriptions of physical specimens

There are very few sightings that could be attributed in all probability to *con rits*. Heuvelmans reports only 20 certain sightings between the 1600's and 1965 for his 'many finned' variety, and the 'diagnostic' criteria employed for this particular creature are applied so liberally, as to be questionable in this case, which may in fact reduce the number of certain sightings considerably and strengthens the case for using the plausibility method in deciding between probable identities for this cryptid.

3.1.1 Encounter number one

Year: 1883.
Location: Hongay beach, Vietnam.
Number of witnesses: Multiple (unknown).

Despite the paucity of high quality sightings, the *con rit* has a most unique qualification as far as sightings are concerned that virtually all other sea serpent types lack, namely there exists a very detailed, visual and tactile description of a beached *con rit* carcass in Hongay, Vietnam made by one Tran Van Con in 1883. Here is the description as related to A. Krempf, Director of the Indo-China Oceanographic and Fisheries Service in 1921 (Heuvelmans, p. 416).

"Here is some information which although it smacks of the marvellous, cannot fail to interest you. I received it at sea from the coxswain of a Customs launch, a 56-year old native called Tran Van Con.

38 years ago (that is to say in 1883, 14 years before Lieutenant Lagrésille's account), this Annamite saw and touched the so-called sea-serpent. Here is the account, faithfully translated: the animal was washed up and dead: it was a carcase in a very advanced state of putrefaction. The head had gone. The body alone was 60 feet long by 3 feet wide.

The animal was formed of successive segments almost all alike one another. Each segment was 2 feet long and 3 feet wide and had a pair of appendages 2 feet 4 inches long.
The teguments were of a remarkable consistency and rang like sheet-metal when hit with a stick. The colour of this tegumentary envelope was dark brown on the dorsal surface and light yellow on the ventral surface.

The stench that arose from this prodigious animal was such that even the Annamites would not go near it, and it was decided to tow the remains out to sea and sink them.

The name given to this animal by my informant is con rit, or 'millipede'. It is this, according to its name and from all the description that I have given you, an Arthropod ... unless it is all a dream, and certainly it is a very detailed, and as another theory about the sea-serpent can do no harm, I have thought fit to send you this information, but ask you to await further details before doing anything about it."

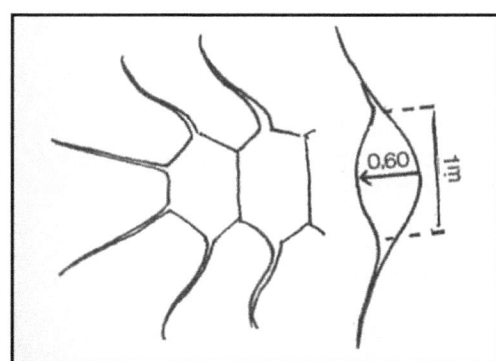

Figure 5: *Krempf's depiction of the con rit's armour, with dimensions.*

Elements of this report were substantiated by a series of follow-up interviews involving Chinese fishermen that were conducted by the immigration authorities in Saigon. The time delay between the sighting and its eventual reporting raises some important questions; such as how accurate can the memory of the details be after all that time, especially given the allegedly highly decomposed nature of the carcass? If validated however (and the broad corroboration amongst the Chinese fishermen goes some way towards this), this report is unique and is certainly detailed enough to serve as a specimen-based description.

3.1.2 Encounter number two

Year: 1899.
Location: Cape Falcon, Algeria.
Number of witnesses: Multiple (unknown).

One other *con rit* sighting of note, was that made in 1899 when the *HMS Narcissus* had rounded Cape Falcon, Algeria. Lieutenant Boothby - who was officer of the watch - observed something unusual in the water at 5 a.m, which he dutifully recorded in the ship's log (Heuvelmans, p. 366).

"Observed a sea monster on the port bow, certainly over 150 feet in length; apparently propelled by large fins, and lying very low in the water."

When a journalist asked a signalman who was also a witness to the creature, if it was not simply a string of porpoises, he gave the following response:

"We saw some porpoises just after and their motion was not the same. You could see the por-

poises jump or tumble over, but this creature lay steadily on the surface, gently gliding through the water ... The monster seemed to be propelled by an immense number of fins. You could see the fins propelling it along at about the same rate as the ship was going. The fins were on both sides, and appeared to be turning over and over. There were fins right down to the tail.

Another curious thing was that it spouted up water [sic] like a whale, only the spouts were very small and came from various parts of the body."

This sighting is remarkable not only because of the degree of corroboration between witnesses, but also due to the fact that the fins were described as being independently movable, which suggests that a multi-limbed organism lies behind the sightings of the *con rit*.

4. SUGGESTING AN IDENTITY

4.1 The case against a cetacean identity

Cryptozoologists freely admit that the classification of the *con rit* is a problem. As has been mentioned, there exist only a handful of sightings of creatures that could plausibly be considered *con rits*. As can be seen from comparing figures one, two, three, and four, the suggested appearance for the *con rit* has changed markedly over the years, yet the view that it is mammalian and very probably a cetacean, has remained largely constant.

The primary problem with the cetacean theory, is that the reconstructions made in recent years by Heuvelmans, and others, have this creature looking completely atypical for a cetacean. The reconstruction by Heuvelmans has the *con rit* somewhat resembling an aquatic ankylosaur (see figure two), replete with bony back plaques and numerous lateral projections. When Heuvelmans was researching and writing *In the Wake of the Sea-Serpents*, the discovery of what appeared to be bony plates in association with fossil archeocetes seemed to lend credence to the idea that they possessed armour-plating of some kind. However this idea quickly fell from favour, and no longer figures in archeocete reconstructions. The notion of lateral fins, although theoretically feasible as they would have aided in locomotion, is also atypical of the cetacea, and unlike the bony plates - have never been proposed in reconstructions of archeocetes. Additionally the fins, which would theoretically be rigid hydroplanes if found in association with cetacea, seem to be contributing to propulsion indicating independent mobility - at least according to the *HMS Narcissus* encounter.

Another major problem with the cetacean identity picture painted by Heuvelmans, Huyghe, Coleman, and Champagne, is one of proportionality. The head, for example, is depicted as being ball-like and located on the end of a thin - but short - neck, which in turn protrudes from a wider set of shoulders; a highly unlikely arrangement for a cetacean. Other features such as a rhomboid or 'delta' shaped bilobate fluke, and terminally located 'nostrils', replete with protruding structures that have been classified as proximate vibrissae, are also characteristics unknown in cetaceans both past and present.

4.1.1 A polychaete identity?

Of the invertebrate theories, Bradshaw's theory that the *con rit* is a polychaete worm is more convincing than the cetacean theories, but has problems of its own. The polychaete are a highly variable class of organisms in terms of their lifestyles and morphology. In addition to which their appearance compliments the general descriptions of *con rits,* as having long tubular bodies with large numbers of protruding filamentous limb-pairs. Also the descriptions of *con rits* as segmented, seem to suggest a certain annelid character. Polychaete worms are also capable of growing to reasonably large sizes - nine feet in the case of the seep-tube worm. However, as was mentioned, there are problems with the polychaete identity. Firstly, the larger polychaetes tend to be at least partial burrowers - this is a protective measure adopted to avoid predation. Burrowing capability might just about seem feasible for a nine feet long worm, but for a 150ft specimen this does not seem plausible. Of course the evolution of armour-plating could be seen as an adaptive compensation to this. However, herein lies yet another problem; namely that no other known representatives of the Annelida have ever developed armour-plating. This weakens the case somewhat for a polychaete identity in the context of a plausibility-method based analysis, as the absence of phylogenetic precedence for a particular inferred trait, increases the need to speculate, which correspondingly decreases the likelihood of that particular trait being plausible.

4.1.2 Or, a Myriapoda identity?

The Myriapoda, like the Annelida, represent a diverse and ancient grouping of organisms. So far, as has been mentioned, few have seriously proposed a Myriapod identity. Researchers have suggested an identity as a member of the Chilopoda, which in order to be plausible would require an evolutionarily improbable scenario involving a member of an exclusively terrestrial class taking to the oceans. Other researchers - falsely citing Dr Karl Shuker - have suggested another possible identity theory; namely that they are large crustacean isopods. This is unlikely as the largest known isopods are not keen swimmers (so are unlikely to be spotted at the surface) and possess only seven pairs of legs (*con rit* witnesses usually report many more pairs of 'fins'). However, it is through an effective fusion of these identity theories that a new identity can be suggested; namely that the *con rit* is a living example of the - currently believed to be extinct - class Arthropleuridea.

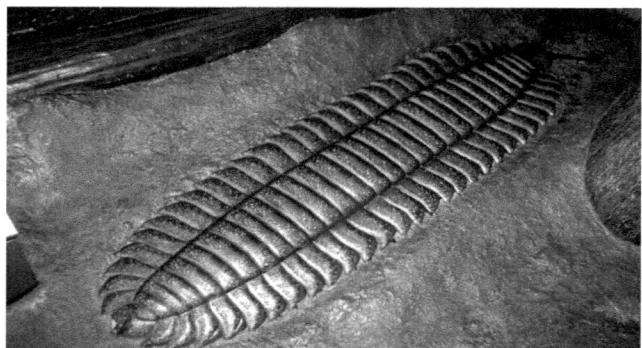

Figure 6. Arthropleura *model from the field museum in Chicago.*

The genus *Arthropleura* contained two known species of Myriapods that became extinct at the beginning of the Permian period. *Arthropleura* was the largest terrestrial invertebrate that ever lived - measuring six to ten feet in length. It was also herbivorous.

4.2 Evolving the *con rit*

One of the main factors that contributed to the extinction of *Arthropleura* was the effect of desertification on its semi-aquatic rain forest habitat. What, if in order to survive, *Arthropleura* took to the seas? There is one main problem with this scenario; namely it is not known if *Arthropleura* had a salt-water tolerance. To address this, it is firstly necessary to point out that the Myriapoda do not seem to be a particularly evolutionarily conservative sub-phylum (as evidenced by the diversity of both extant and extinct body plans); opening the possibility that it could have evolved a tolerance relatively rapidly. Also it is worth noting that *Arthropleura* is thought to have evolved from crustacean-like ancestors in the Carboniferous, which may have meant that it managed to retain some of its marine adaptations. *Arthropleura* is in fact thought to have had an amphibious lifestyle; perhaps it retained a tolerance for salt water also? Speculating on an evolutionary narrative to explain its transition from an amphibious to an exclusively marine lifestyle is not difficult, if we permit one of the aforementioned scenarios.

The creeping desertification of the land, would have destroyed its terrestrial food sources, resulting in a strong selective pressure in favour of marine habitat exclusivity. *Arthropleura* is thought to represent the absolute upper size-limit for a terrestrial invertebrate in an abnormally oxygen-rich environment. However, these restrictions are overcome in a marine environment, where hydrostatic support provided by water coupled with an absence of major predators, and the effects of delayed sexual maturity as an adaptation to scarcer deep-sea food sources (deep-sea gigantism) would have theoretically allowed it to grow to considerable sizes.

As was mentioned, this scenario is effectively a fusion of the chilopod Myriapoda, and Shuker's mystery crustacean identity theories, as it posits that an amphibious early myriapod with crustacean characteristics took to the oceans as an adaptation to early Permian habitat destruction through desertification.

4.3 Does it fit with the facts?

Con rits are rarely seen, which would seem to indicate that they only very rarely surface, which is compatible with the idea that - like the giant squid - they are huge invertebrates adapted to life at abyssal depths. It is possible that they have also adapted to carnivory, feeding on organisms such as deep-sea sharks, squid, and maybe even small to medium sized whales.

Bradshaw has proposed an interesting theory regarding the apparent 'walrus' like head that has been described for this cryptid. He suggests that the *con rit* could possess fangs similar to those found in the smaller polychaetes, which could - from a certain perspective - look like tusks. The same could be true of the *con rit* if it were a giant myriapod, as many myriapods including *Arthropleura* possessed fangs, although it is not thought that *Arthropleura* had a venomous bite. Perhaps as an adaptation to a carnivorous lifestyle, like its terrestrial cousins, the *con rit* acquired the ability to envenomate prey at some point in its evolution.

4.4 A need for reclassification

So far, only Heuvelmans has seriously suggested a parataxonomic name for the *con rit*. However, the classificational nomenclature he chose to employ, was biased by his assumption that the *con rit* was an archeocete - as is evidenced by the name he chose: *Cetioscolpenda aelani* (Heuvelmans, 1965) - literally 'Aelian's cetacean centipede'.

This author has suggested that the data strongly tend to support an invertebrate identity for the *con rit*, and that it should therefore be reclassified into the newly proposed genus *Mariascolpenda* (Sea centipede).

Its parataxonomic name in full could read in full as *Mariascolpenda aelani* (Woodley, 2007?), comprising a new family and a new order in the class Arthropleuridea in the sub-phylum Myriapoda.

This author *further* suggests, that the specimen-based description of the beached carcass observed by Tran Van Con in 1883, and described to A. Krempf in 1921 may represent sufficient in-hand evidence to constitute a type specimen for this claim.

The super-otter and the many-humped sea serpent: close cousins?

1. INTRODUCTION

1.1 Overview

The first technical description of the hypothetical super-otter was made by Heuvelmans, who suggested it as an identifier for a group of cryptids that were commonly being seen in the cold northern oceans of the Arctic up until 1848 - when sightings suddenly ceased.

Christening it with the parataxonomic name of *Hyperhydra egedei* (literally; 'Egede's super-otter'), this cryptid was the putative identity of the 'Great sea serpent' as witnessed by Hans Egede, Bishop of Greenland, in 1734.

The many-humped sea serpent, to which Heuvelmans gave the name of *Plurigibbosus novae-angliae* (literally; 'That with many humps from New England'), seems to be a very similar cryptid to the super-otter, in terms of both inferred broad ecological range and general characteristics. However, Heuvelmans insisted on the idea that they were only distantly related at best, despite the fact that he identified both cryptids as archaeocetes.

1.2 Descriptions

As was mentioned, both the super-otter and the many-humped sea serpent, are similar in appearance; however there are certain key anatomical and ecological points on which they are said to differ.

1.3 The super-otter

Described as being between 65 and 100 feet in length, this creature is said to have a longish neck, four webbed feet, and a long tail that undulates as the cryptid moves its body whilst swimming; giving rise to a 'multiple humped' profile with six to seven 'humps' visible at a time. It is said to possess skin of a pale brown colour, with a rough and wrinkled texture. Intriguingly, up close, the skin is described as being covered in a fine layer of woolly hair.
The super-otter's head is said to be flattened on the top, and tapering. Its eyes are described as being small. However, its teeth are said to be visible when the cryptid opens its mouth.
The super-otter's range is quite restricted, being confined to a probable pelagic habitat within the Arctic Circle north of Norway, and south of Greenland. Migratory patterns are either un-

clear, or absent from the data. Heuvelmans postulated that it may move south during the summer to breed.

Heuvelmans has suggested that the super-otter might now be extinct due to possible competition with long necked seals of the genus *Megalotaria* (see figure fourteen, previous chapter) as sightings of this creature have apparently not been made since the middle of the nineteenth century.

Figure 1. *The super-otter by Cameron McCormick (2006).*

1.4 The many-humped sea serpent

Described as being in basically the same size range to the super-otter (60 to 100 feet), this cryptid is also supposed to possess a moderately long neck. However, where it differs from the super-otter, is in the fact that it seems more adapted to life in the oceans, as it is said to possess a cetacean-like bilobate fluke, and flipper-like forelimbs. It is also described as being capable of attaining speeds of 22-40 knots.

The many-humped sea serpent moves in an undulating manner similar to the super-otter, which partly gives rise to its 'many-humped' profile. However, it is also described as possessing a series of static 'humps' arranged along its back, whose suggested functions range from the idea that they are stores for additional air used to prolong dives, to Ivan Sanderson's suggestion that they are in fact inflatable ballast organs used for stability.

The cryptid is often described as bi-coloured, being dark brown on its dorsal side, and paler on its ventral side. Its skin is said to be smooth, although according to Heuvelmans, some reports have described it as being 'scaly' possibly due to the presence of marine parasites. Another interesting feature, is the presence of a triangular dorsal fin, which Heuvelmans suggests is a dimorphic characteristic present in some males only.

The many-humped sea serpent seems to have a similar niche to the super-otter, however its range appears to be broader and more southern, with a distribution that encompasses the coastal UK, southwest Iceland, the New England coast, and the Gulf of Mexico. Sightings data from this latter part of its range, seem to strongly imply that during the winter months the many-humped sea serpent migrates southward.

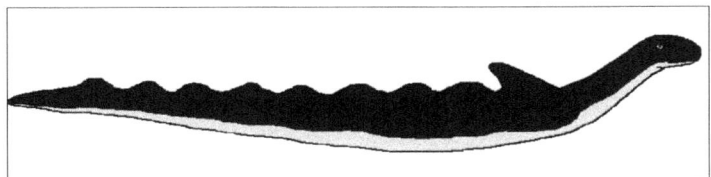

Figure 2. *The many-humped sea serpent by Cameron McCormick (2006).*

2. IDENTITY THEORIES

2.1 Heuvelmans' proposed identities

Heuvelmans suggested that both the super-otter and the many-humped sea serpent were cetaceans belonging to the ancient suborder archaeoceti, although, as has already been mentioned, he believed them to be only distantly related to one another at best. The primary traits, which Heuvelmans employed in suggesting this identity, were their otter-like necks. Both cryptids are described as possessing moderately long necks - a trait that no modern cetacean is known to possess. However, Heuvelmans predicted that early examples of the archaeoceti might have possessed this (and other) otter-like characteristic. It was from this that the equivalent characteristics in these two hypothetical archaeocetes were supposed to be derived.

Heuvelmans intended his three archaeocetes to be representative of cetaceans at different stages of their evolution, and confined to different, but overlapping, niches. The super-otter was to be the most primitive of the three, retaining many terrestrial adaptations, and even possessing the ability for terrestrial locomotion, whilst being confined to the northern waters of the Arctic Ocean. The many-humped sea serpent was supposed to be a representative variant of early aquatic archaeocetes, occupying a broader and more southerly niche than the super-otter, whilst the many-finned sea serpent was supposed to represent a more specialized variety of basilosaur, equipped with armour, and restricted to the equatorial regions.

As an afterthought, Heuvelmans also suggested a possible sirenian identity, as an alternative to the primitive archaeocete identity for his super-otter category.

2.1.1 Is it consistent?

As was illustrated in the previous section, a cetacean identity did not seem appropriate for the many-finned sea serpent, or *con rit*, where, it was argued a myriapod identity better fits the observational data. Although in the case of the *con rit,* an archaeocete identity had been proposed centuries before Heuvelmans. The eagerness that he exhibits in assigning these other two (unrelated) cryptids to the same category, seems to imply that the label of 'archaeocete' is employed as more of a catch-all term for these more ambiguous and hard to classify cryptids, rather than as something that is necessarily a probable indicator of their true identities *per se*.

As for a sirenian identity for the super-otter, this can safely be ruled-out on the basis that there are no sirenians with anything other than vestigial pelvises, or long necks. Either the super-otter evolved away from other sirenians in this respect, or as is more likely the case, it is related to something, which *already* has a longish neck and long tail.

2.1.2 Archaeocetes

As has been mentioned, Heuvelmans predicted that the early examples of the archaeoceti might have had certain otter-like characteristics. It is true that some examples of transitional forms had amphibious characteristics; such as *Ambulocetus* - discovered in Pakistan by the anatomist and palaeontologist Johannes Thewissen, which had the ability to walk on land, as well as to exploit both fresh and salt water environments.

Figure 3. Ambulocetus.

However, in terms of general morphology, as can be seen from figure three, *Ambulocetus* more closely resembles a mammalian crocodile than an otter; with its truncated head, and row upon row of sharp teeth adapted for the lifestyle of an ambush predator. Probably the only archaeocete that bears any resemblance to Heuvalmans' super-otter, or his many-humped sea serpent, is the remingtonocetid *Kutchicetus*, which was the size of a river otter, and lived during the early Eocene.

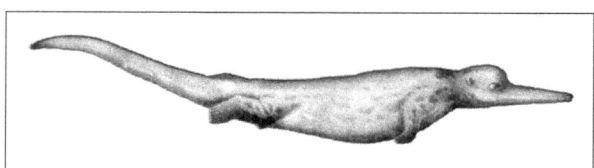

Figure 4. Kutchicetus.

Whale evolution has been characterized by a move towards greater degrees of streamlining, with a reduction in the amount of body hair, a reduction in the size of the hind limbs, and the reduction in neck length, all of which cause unnecessary drag. For the super-otter, or the many humped sea serpent, to look the way that they do, traits that were specifically selected against in the early cetaceans for the purposes of greater adaptation to an exclusively aquatic lifestyle (such as neck length, fur, and bodily 'lumpiness') would have to have been selected for, and amplified. The presence of these non-cetacean traits in these cryptids does not necessarily rule out the possibility that they exist; it simply implies that an archaeocete identity for these cryptids is a very poor match.

The only logical identity theory for these cryptids, is as members of the lutrinae sub-family - the otters. As will be demonstrated subsequently, the traits of these cryptids make them a remarkably good match for the lutrinae, and it is indeed surprising that Heuvelmans so readily dismissed this potential identity theory.

3. SUPER-OTTERS AND MANY HUMPED SEA SERPENTS AS LUTRINAE

3.1 Overview

The lutrinae are a sub-family of amphibious (and in one case aquatic) carnivorous mammals comprised of thirteen species in seven genera. Fish are their primary food source; however this may on occasion be supplemented with other animals, including crustaceans and molluscs, depending on the species.

Size-wise, most members of the lutrinae are small, being under a metre in length on average; however two species can reach relatively large sizes - the sea otter and the giant-otter, both of which can easily reach up to six feet in length.

Otters are characterized by the presence of a thick muscular tail, which makes up 30% of their body length (the exception to this is the sea otter). They are also highly social animals, especially in the case of the sea otters, which employ stones for the task of breaking open mollusc shells, exhibiting one of the few instances of non-Human tool use in the animal kingdom.

3.2 More alike than different

Although Heuvelmans claimed that the super-otter, and the many-humped sea serpent, are probably only very distantly related, a strong case seems to present itself for their being probably quite *closely* related.

Both are very similar in profile to one another, and both have been sighted in similar locales. In attributing a particular sighting to one or the other category, it does seem that Heuvelmans is pigeonholing to a degree. However, it also seems that Heuvelmans has built a sufficiently strong case for these cryptids being different on certain key points of anatomy, such as colouration, and the presence of flippers vs. webbed feet etc.

3.2.1 A lutrinae identity for the super-otter

The super-otters are probably the less ambiguous of the two cryptids as far as inferring a lutrinae identity is concerned. Firstly, it is described as possessing a highly flexible body, which interestingly enough, is a characteristic most pronounced in the aquatic sea otters. It has a very long tail, along which undulations propagate giving rise to the many 'humps' that witnesses describe - something which is also observed in the long-tailed otter species, such as the giant otter. It has also been reported as having a covering of woolly hair - an insulating trait that would be expected of the super-otter, if it were - in fact - a member of the lutrinae. It is difficult to know whether or not the super-otters are amphibious, or largely aquatic. In all likelihood it is the latter, and in that respect shares this trait in common with its sea otter cousins. After all, a member of the lutrinae of such vast proportions, and weighing many tons is going to have considerable difficulty manoeuvring on land or ice.

3.2.2 A lutrinae identity for the many-humped sea serpent

This is the more ambiguous of the two, as certain traits that it possesses seem to be compatible with a lutrinae identity, whereas other traits are completely incompatible. The bi-colouration that it exhibits, with the darker brown being displayed on its dorsal side, and the lighter brown on its ventral side, seems in strong agreement with a lutrinae identity, as many otters do in fact display this very same combination of colours.

Figure 5. A European otter exhibiting marked bi-colouration.

The presence of cetacean-like fins instead of webbed feet, is - however - more difficult to reconcile. One possible explanation is to look at the sea otters, which - with respect to their adaptations to a wholly aquatic lifestyle - are the most advanced of the lutrinae. These otters have feet, which have been fused into pseudo-flippers; a trait that makes access to land very difficult, but has allowed them to capitalize on their aquatic lifestyle. The bilobate tail, which has been attributed to them, could simply be the product of a hypothetical fusion of the hind limbs and the reduction in tail length, as with the pinnipeds. It is interesting to note that tail proportions are also greatly reduced in the sea otters, probably as an adaptation to their lifestyles. If the many-humped sea serpent simply represents a highly advanced form of lutrinae, then this trait could simply be the logical outcome of an adaptation to an exclusively aquatic lifestyle.

The three traits that are the most difficult to reconcile with a lutrinae identity, are the presence of the 'static' humps along the back of the cryptid, the apparent absence of any fur, and the presence of a dorsal fin. The 'static' humps, which as was mentioned have had a variety of functions attributed to them by both Heuvelmans and Sanderson, could simply be undulations propagating down the great length of their highly flexible spines - as in the case of the super-otter. The other aforementioned proposed functions such as the idea that they represent ballast organs or air storage organs for prolonging dives, seem to be more convoluted explanations, and are simply less likely to be the case.

The supposed absence of fur, or the presence of smooth or 'scaly' skin in the many humped sea serpent, could perhaps be due to the secretion of large amounts of squalene (a sebaceous gland derived insulating chemical) coupled with the development of a more rubbery skin, and larger deposits of subcutaneous fat, as a way of overcoming the need to be hairy. The bi-colouration in this case, could be due to pigment in the skin, as in the case of the orcas. A more plausible explanation is that these cryptids are in fact hairy, like their super-otter cousins; it is just that those who have been lucky enough to witness them, simply haven't been close enough to see any hair, which may also be slicked back against the cryptid's skin, thus giving the erroneous impression of a smooth, hairless body.

The dorsal fin seems to be an anomaly. The fact that it is a trait only associated with certain sightings led Heuvelmans to suggest that it was a dimorphic trait that was only to be found on the largest males. It is far more likely that due to its odd inconsistency (there are no other significant points of correlation between sightings of 'dorsal finned' many-humped sea serpents), this trait is a simple misidentification of something else, and can be safely dismissed on that basis.

3.3 Their standing within the sub-family lutrinae

A comparative analysis of the two cryptids with other, known, members of the lutrinae would seem to suggest that Heuvelmans placing them into separate, new genera is appropriate given a lutrinae identity. Whereas the super-otters possess a more 'traditional' body plan, the many-humped sea serpent would seem to be the more advanced of the two in terms of perceived adaptations to an exclusively aquatic lifestyle. This would mean that the super-otter genus *Hyperhydra* is likely to be allied with the otters in the genus *Lutra,* given their mostly northern European range. Whereas the many-humped sea serpent genus *Plurigibbosus,* may possibly be allied with the sea otter genus, *Enhydra,* although, given the sea otter's exclusively Pacific distribution and the many-humped sea serpent's exclusively Sub-Arctic/Atlantic/Gulf of Mexico distribution, it may be the case that *Plurigibbosus* and *Enhydra* are less closely related genera than *Hyperhydra* and *Lutra.*

4. SIGHTINGS

4.1 Overview

Although not intended as a complete review of the sightings data available on these cryptids, in this section the first sightings of both of these cryptids (from which they derived their para-taxonomic names) will be described, and the attendant difficulties in analysing them will be briefly explored.

4.2 Encounter number one

Year: 1741
Location: North Atlantic, south of Greenland.
Number of witnesses: Two, possibly more.

Hans Egede was the founder of a protestant missionary colony in Greenland, for which he became widely known as the Apostle of Greenland. Whilst on a ship off of the southern coast of Greenland, he was witness to a most unusual sight. The following description can be found in both the 1745 English translation of his book, *A Description of Greenland* and in Heuvelmans (p. 100).

"As for other Sea Monsters...none of them have been seen by us, or any of our Time, that ever I could hear, save that most dreadful Monster, that showed itself upon the Surface of the Water in the Year 1734, off our New Colony in 64 Degrees. This Monster was of so huge a Size, that coming out of the Water, its Head reached as high as the Mast-Head; its Body was bulky as the Ship, and three or four times as long. It had a long pointed Snout, and spouted like a Whale-Fish; great broad Paws, and the Body seemed covered with shell-work, its skin very rugged and uneven. The under Part of its Body was shaped like an enormous huge Serpent, and when it dived again under Water, it plunged backwards into the Sea, and so raised its Tail aloft, which seemed a whole Ship's Length distant from the bulkiest part of its body."

Figure 6*. Hans Egede's sea serpent.*

Heuvelmans seems convinced that Egede counts as a valid witness. His writing is described as being *"meticulous, sober and rather dry"*, with the absence of 'exaggerating' and 'flowery' prose seeming to count in his favour. The debate, however, is not over whether or not Egede could have made up his sighting; it is over exactly what it is that he saw. Charles Paxton, who was mentioned briefly in chapter one, formulated the hypothesis that Egede actually saw a whale's penis in a state of arousal. Although it is an interesting hypothesis, and could certainly account for some sea serpent sightings, it is ambiguous as to how well it fits the facts in this case. The scallop patternation that Egede describes on the body of the creature that he saw, is consistent with that found on certain cetacean penises; however the size of the creature observed by Egede, and the presence of facial characteristics are clearly inconsistent with this hypothesis. For the Egede cryptid to be a whale penis, serious exaggeration has to be assumed on the part of Egede. As Heuvelmans points out, Egede's writing style is too factual and objective. Additionally, his son, who was also a witness to the event, corroborated much of his report. Could Egede have been one of only a handful of witnesses to a sea serpent of the super-otter variety? Possibly, however simply because the cetacean penis identity does not seem a likely culprit for his 'Most dreadful monster', does not mean that he was necessarily witness to something new to science.

4.3 Encounter number two

Year: 1639.
Location: Cape Ann.
Number of sightings: Multiple (unknown).

A variety of sightings of sea serpents were made off of the New England coast between the seventeenth and nineteenth centuries. The descriptions are broadly convergent on details like size, the presence of multiple humps, and an elevated head. The first sighting (by westerners at any rate) was made in 1639, and was related to the naturalist John Josselyn, who recorded it in his 1674 book, *An Account of Two Voyages to New England*. The account also appears in Heuvelmans (p. 143).

"They told me [in 1674] of a sea-serpent or snake, that lay coiled upon a rock at Cape Ann; a boat passing by with Englishmen on board, and two Indians, they would have shot the serpent, but the Indians warned them, saying that if he were not killed outright, they would all be in danger of their lives."

Although not a first hand account, this sighting compares favourably with many others made in and around the area over the course of the next few centuries. This type of sea serpent was even 'formally' recognized with the scientific name of *Scoliophis atlanticus* by the New England Linnaean Society in 1817, after the discovery of what turned out to be little more than a deformed land-snake.

This set back aside, it is interesting to note that many (although not all) of the east coast sea serpent sightings were classified by Heuvelmans as belonging to his many-humped category, yet in general these east coast sea serpents seem to have large mobile humps, and although they differ from the 'pawed' super-otters, they seem to bear little resemblance to the 'static humped' many-humped sea serpent of Heuvelmans. This lends added credence to the idea that the many-humped category simply needs to be reclassified, and a more detailed and discriminatory study needs to be made with a view to disentangling the various sightings that Heuvelmans places in his 'many-humped' category.

Figure 7. The Cape Ann sea serpent.

5. CONCLUSIONS

5.1 A third lutrinae cryptid?

There exist reports of a cryptid native to the west coast of Ireland known as the master-otter or in the local vernacular as *dobhar-chu*. This large otter-like creature is described as being several times the size of a grown man, and highly aggressive, having reputedly killed a girl by the name of Grace Connolly in 1722, who lived near to Glenade Lake in County Leitrim, wherein the creature was said to reside. Her husband apparently killed the creature, whose partner became enraged and attacked him. The second creature was apparently also killed. Her tombstone, which is dated 24th of September 1722, has inscribed onto it an image of a large otter-like animal, complete with a long tail possessing a tufted tip, a long neck, and a short head with ears. Sightings of a creature matching this description off the coast of County Mayo were apparently being made as late as 1968.

It seems that the master-otter could be a smaller, more amphibious, relative of the many-humped sea serpent. The location of the sightings seems to indicate that it shares some of its range with its larger, fully aquatic, lutrinae cousin.

5.2 Summary

The many-humped sea serpent seems to have reasonable numbers of sightings in its favour, whereas the sightings data for the super-otter seems to be scarcer. According to Heuvelmans, the many-humped sea serpent has 59 total and 33 certain sightings, whereas the super-otter has only 28 total and 13 certain sightings. There is less data available on these cryptids than are available for Heuvelmans' pinnipeds, however - comparatively speaking - these speculative lutrinae have a much more restricted range than long necked pinnipeds, so are correspondingly less likely to be observed.

The Paxton paper, which made a convincing case for the Hans Egede sighting being a whale in a state of arousal, seems to pose a strong challenge to the potential reality of long bodied, multiple humped cryptids. As was mentioned previously, Paxton's 'cetacean arousal hypothesis' could be used to explain a variety of sightings. However even if the 'cetacean arousal hypothesis' can be used to satisfactorily dismiss the Egede account, does it really suffice as a universal explanation for sea serpent sightings of this type? Can it for example account for why many sightings of these long bodied, multiple humped sea serpents clearly describe the presence of eyes? Or woolly fur? Or even a clearly defined mouth, and set of teeth? Obviously a case can be made for human biases coming into play. After all, it is certainly the case that human psychology is in the habit of anthropomorphizing and inferring features in things that do not really have those features (seeing animal shapes in clouds etc.), but then it is necessary to consider the fact that many of the features observed correlate across broad sweeps of time and geography. Is it simply the case that unrelated witnesses are sharing a common set of delusions? Or is the coincidence evidence that they are witnessing marine organisms, possibly undiscovered members of the lutrinae sub-family, which have successfully eluded the attentions of zoologists?

CHAPTER 3.
The Others

· Marine saurian

· The Super-eel

· Giant invertebrate

Marine 'saurians':
genuine archaeocetes?

1. INTRODUCTION

1.1 Overview

The 'marine saurian' as Heuvelmans described it, is remarkable for a variety of reasons, not least of which is due to the offhanded manner in which Heuvelmans treats it. No attempt is made to illustrate it, nor were any bold suggestions forthcoming on parataxonomic names. This cryptid is simply described somewhat dismissively as being either a thalattosuchian crocodile or a mosasaur, without much more thought being given to the matter than that. Heuvelmans' views on this cryptid were obviously based on the idea that if it *looks* reptilian then it must *be* reptilian, and the only animals corresponding to the observed characteristics and proportions of the marine 'saurian' that Heuvelmans would have known about in the 60s would have been thalattosuchian crocodiles or mosasaurs and their known allies. It will be argued that a saurian identity is not the most probable or logical for this cryptid, and that recent discoveries of new transitional fossils in the evolution of whales, yield a much more plausible archaeocete identity for this category.

1.2 Description

This cryptid is usually likened to a crocodile by those who see it. It is described as being in the range of 50-60 feet in length, with a long head filled with closely set teeth. Its tail is said to be long, making up a considerable amount of the cryptid's overall body length. It has been described by some witnesses as being able to move its body in the horizontal plane, and possessing smooth, reddish brown skin, with few or no visible scales. The distribution of sightings data indicate that this cryptid is mildly cosmopolitan, and pelagic for the most part, but prefers the warm water of the equatorial regions as most sightings seem to have been made there. Although interestingly enough, a sighting of this creature was made in the cold waters of the North Sea. Its two pairs of limbs are often described as being 'webbed', and sometimes a dorsal ridge is described as running along its back.

Figure 1: *A generic marine saurian by Cameron McCormick (2006).*

2. IDENTITY THEORIES

2.1 Thalattosuchians and mosasaurs

As has been mentioned, Heuvelmans advocated the theory that the animals classed in his 'marine saurians' category were relict populations of giant marine reptiles, currently thought to be extinct. There are a variety of such reptiles, including ancient crocodiles, that have at one point or another in Earth's past assumed the basic crocodile body plan. These include:

2.1.1 Thalattosuchia

These enormous crocodiles lived from the end of the Jurassic into the Cretaceous period. They are most remarkable for being strongly adapted to an exclusively marine existence. In the current era, only one species of crocodile has demonstrated an affinity for persistent exposure to marine environments; the double crested crocodile. This 20-feet long native of the waters of Indonesia and Northern Australia is still predominately adapted for an amphibious existence, as opposed to an aquatic one, whereas the thalattosuchia possessed feet, which had been transformed into flippers and an ichthyosaur-like tail fin, which would have been its primary organ of locomotion.

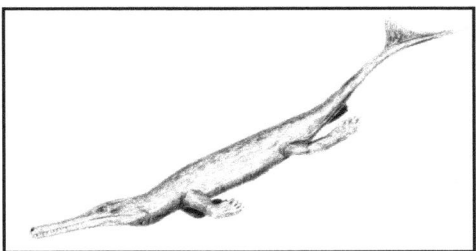

Figure 2: <u>Teleidosaurus</u>; *an extinct thalattosuchian crocodile.*

2.1.2 Mosasaurs

Mosasaurus means 'lizard of the Meuse', as it was from this river in the Netherlands that the first fossil specimens were recovered. Living around 70 – 65 million years ago during the Cretaceous period, they were generally similar in appearance to the thalattosuchian crocodiles, although they lacked the modified tail. Their tails however are thought to have factored into their propulsion quite significantly, as they were long, muscular, and relatively flattened in the vertical plane. Despite superficial similarities, the mosasauria were largely unrelated to the thalattosuchia and modern crocodiles. As a matter of fact these animals are thought to be more closely related to snakes and monitor lizards. The similarity between mosasaurs and the thalattosuchia is undoubtedly due to convergent evolution, as both are thought to have occupied very similar niches.

Figure 3: Mosasaurus.

2.2 Crocodile-like archaeocetes

There are a variety of early archaeocetes that resemble crocodiles not just in form, but to a degree, in function also. Take the primitive *Ambulocetus* from the Eocene, which was mentioned briefly in the section on the super-otter and the many-humped sea serpent. This primitive archeocete was almost certainly an ambush predator, which had adopted, to a certain extent, the lifestyle of a crocodile, and in doing so had come to resemble one in form. Even more remarkable is *Rodhocetus*, which lived approximately 47 million years ago. This primitive whale was even more strongly saurian in form, complete with an elongated, crocodile-like skull, four flippers, and a tail (although one that was proportionally diminished in size when compared to either the thalattosuchia or the mosasaurs) which may - according to some reconstructions - like that of the thalattosuchia, have ended in a special fluke to assist in propulsion.

The basilosaurs, which lived 40 to 34 million years ago, and were discussed briefly in the first chapter, also possess certain crocodile-like characteristics, including an elongated skull filled with sharp teeth, and a long, sinuous body.

Figure 4: Rodhocetus.

2.3 Comparisons

Heuvelmans based his reptilian identification for the marine saurians on two primary observations. Firstly, he suggested that as mosasaurs were predators well suited to making deep dives, they could quite effectively remain unseen. Secondly, the sightings, as were mentioned in the introduction, seem to place this cryptid in the tropical regions, which again would support the notion of them being reptilian. However, many factors would seem to point strongly towards an archaeocete identity. The question has to be raised as to what degree the inferred traits of Heuvelmans' marine saurians, correspond with the traits in other non-reptilian animals? This has been addressed already, with the conclusion that many examples of the early archaeoceta exhibited crocodile and mosasaur-like traits. Especially interesting in this regard are the basilosaurs, which this author believes to be behind the majority of the so-called marine saurian sightings, as they possess some remarkably convergent traits with the mosasauria and similar large marine reptiles. These traits, in addition to size and general proportions also include the presence of a dorsal ridge and interestingly enough the ability to move effectively in a *horizontal* anguilliform (eel-like) motion in addition to a vertical one - a trait which is extraordinarily reptilian and has no known precedent or counterpart in the cetaceans.

It is evident then that the basilosaurs satisfy nearly all of the requirements for Heuvelmans' marine saurians, however there are yet other reasons for suspecting an archaeocete identity. Firstly there is the 1998 paper of Charles Paxton in which he introduces his famous cumulative distribution curve as a metric of what there is left in the way of long bodied marine organisms to discover. The conclusion that Paxton arrived at from this exercise was (as was mentioned in chapter one) that the majority (but not all) of the large bodied species that are as yet to be identified are probably cetaceans. This is based on the observation that when a new species is discovered, it is usually a cetacean rather than anything else, which implies that cetaceans make up the majority of all unknown, large bodied marine organisms to be discovered.

Secondly, there is simply the issue of `Occam's Razor`, as thalattosuchian crocodiles and mosasaurs both became extinct around 70 million years ago during the Cretaceous, whereas the basilosaurs became extinct comparatively much more recently, around 40 million years ago. Of the two, the basilosaurs are the more likely to have survived, not only due to the fact that it is a cetacean and cetaceans are the dominant large marine predators of the current age, but also because it evolved after the KT extinction event whereas the thalattosuchian crocodiles and mosasaurs seemed to have been made extinct because of it. To suggest a prehistoric reptile identity for the marine saurian category, would not only require that a case be made for long term survival, but would also require that one be made for *why* these large reptiles survived the KT extinction event, which simply adds to their improbability as a candidate identity for the sightings that Heuvelmans placed within this category.

3. SIGHTINGS

3.1 Overview

Heuvelmans records relatively few sightings of his 'marine saurian' (only nine in total, of which only four are of high quality), however the sightings are all highly consistent with one another, which adds credibility to the idea that there is something 'new' being observed. Some of the accounts of this cryptid are highly interesting in terms of what is being described, as the opportunity to make fairly detailed observations seemed to present its self in a quite a few of the sightings.

3.2 Encounter number one

Year: 1877.
Location: Latitude 31° 59' N, longitude 37° W (mid-Atlantic).
Number of witnesses: Two.

On the 30th of July 1877, John Hart, the helmsman of the *Sacramento*, called the ships commander, Captain W. H. Nelson, on deck claiming to have seen a sea serpent. The captain, being naturally sceptical of this claim, procrastinated before eventually arriving at the scene only to witness the monster moving away from the ship. On the 20th of October the *Sacramento* docked in Melbourne, and on the 24th of November the helmsman gave the following description of the creature to the *Australian Sketcher,* which was accompanied by a drawing of the creature.

The full account can be found in Heuvelmans (p. 277).

"It had the body of a very large snake; its length appeared to me to be about fifty feet or sixty feet. Its head was like an alligator's, with a pair of flippers about ten feet from its head. The colour was a reddish brown. At the time seen it was lying perfectly still, with its head raised about three feet above the surface of the sea, and as it got thirty or forty feet astern, it dropped its head."

This particular account seems honest enough; there are two witnesses who independently corroborated elements of the story, and both witnesses are ranking officers aboard a ship, which gives them professional credibility. It is possible that *both* could have collaborated on the creation of this story, but for what gain? Temporary notoriety? To see their names in the papers? It is far more likely that the captain, especially - by lending his name to this story - would have put his own reputation at risk; therefore seeing something genuine might have compelled him to take this risk, and report what he saw truthfully. Unless there are details of this story unknown to this author (like evidence of untrustworthiness on the part of one - or more - of the witnesses) then it is probably safe to conclude that Hart and Nelson were witnesses to *something* unusual in the waters of the mid-Atlantic on that day.

Figure 5: *The Sacramento sea serpent (1877).*

3.3 Encounter number two

Year: 1915 .
Location: North Atlantic.
Number of witnesses: Six.

In 1915, the German submarine, *U 28*, was on patrol in the North Atlantic when it spied the British steamer *Iberian*, however when it successfully torpedoed and sank the ship, something quite unexpected came to the surface. The U-boat commander, Georg Günther Freiherr von Forstner recounted the incident in a newspaper. The article can be found in Heuvelmans (p. 395 - 396).

"On 30 July 1915 our U 28 torpedoed the British steamer Iberian (5,223 tons) carrying a rich cargo in the North Atlantic. The steamer, which was about 600 feet long, sank quickly, the bow sticking almost vertically into the air, towards the bottom a thousand fathoms or more below. When the steamer had been gone for around 25 seconds, there was a violent explosion at a depth which it was clearly impossible for us to know, but which we can reckon, without risking being far out, at about 500 fathoms. A little later pieces of wreckage, and among them a gigantic sea-animal, writhing and struggling wildly, were shot out of the water to a height of 60 to 100 feet.

At that moment I had with me in the conning tower my officers of the watch, the chief engineer, the navigator, and the helmsman. Simultaneously we all drew one another's attention to this wonder of the seas. As it was not in Brockhaus nor in Brehm we were, alas, unable to identify it. We did not have the time to take a photograph, for the animal sank out of sight after 10 or 15 seconds...It was about 60 feet long, was like a crocodile in shape and had four limbs with powerful webbed feet and a long tail tapering to a point.

That the animal should have been driven up from a great depth seemed to me very understandable. After the explosion, however it was caused, the 'under-water crocodile', as we called it, was shot upwards by the terrific pressure until it leapt out of the water gasping and terrified."

A most remarkable account. However it suffers from some factual inconsistencies; as Heuvelmans points out, the wreck would have to have been sinking at a speed of around 90 miles per hour, to have descended 500 fathoms in 25 seconds. The ship was probably not far below the surface when it detonated, as was the cryptid, which would account for why it 'shot' out of the water like a champagne cork, riding the shock wave. As with the *Sacramento* sighting, the witnesses are impeccable in terms of their professional credibility. It is highly unlikely that they would have conspired to invent such a story, although a case could always be made suggesting that they created it to embellish an already remarkable tale (namely the sinking of a British ship that was obviously packed with high explosives).

3.4 Encounter number three

Year: 1983.
Location: Bungalow beach, The Gambia.
Number of witnesses: Numerous (unknown).

Fifteen year-old Owen Burnham was out walking along the beach with his family, when they came across the beached carcass of what is referred to now in cryptozoological circles as 'Gambo'. Burnham, who was a wildlife enthusiast, insisted on measuring it and making sketches. He made no effort to take any samples, as it was only after he had gone home that he realized that he could not identify the carcass from books. Local villagers described the remains as those of a dolphin, but Burnham's sketches reveal only a superficial similarity. Burnham described the 'Gambo' carcass as being relatively free of decay, around fifteen feet in length with a brown dorsal colouration and a white ventral colouration. The body itself was around six feet in length, with a five foot girth. He took detailed measurements of its domed head, which was ten inches tall, one foot wide, and was a total length of 4.5 feet. The head possessed a beak, which was on the order of 2.5 feet in length, 5.5 inches tall, and five inches wide, with a small pair of nostrils at the tip. It possessed 80 uniform, conical teeth, and was described as having small eyes.

The carcass supposedly had two pairs of flippers - one front pair which measured 1.5 feet, by eight inches each - and a rear pair, of which one was nearly dismembered, revealing a portion of intestine. The tail, interestingly enough, was around five feet in length, and tapered to a point, showing no evidence of a fluke, and there was no evidence of a dorsal fin.

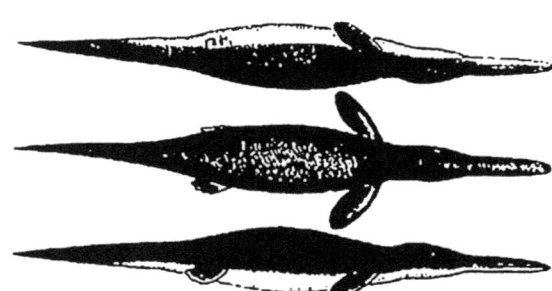

Figure 6: The 'Gambo' carcass reconstructed from illustrations by Owen Burnham by Mark North (2006).

The 'Gambo' mystery carcass has generated considerable debate in cryptozoological circles regarding its putative identity, or even its existence. Darren Naish finds the fact that no samples were taken to be highly suspicious, and possible evidence that the story is a fiction. Cryptozoologist Chris Orrick believes it to be the remains of a Shepherd's Beaked Whale, mangled so that the genital slit and dorsal fin were aligned in such a way as to give the impression of a torn limb. Karl Shuker has noted that it appears to have similarities with a pliosaur, a thalattosuchian crocodile, and a mosasaur, but in correspondence with the CFZ has stated "for the simple reason that without a carcase or at least some tissue samples to analyse, it is impossible to state with certainty what it is/was".

The Centre for Fortean Zoology even launched an expedition into The Gambia to try and unearth the remains of 'Gambo' (which had been buried after having had its head sold to a tourist reputedly) but could locate nothing.

The existence of the carcass seems highly likely, and Burnham's reasons for not taking a sample also seem plausible (not only did he not know that he had stumbled upon something controversial, but he might also have encountered strong resistance from his family if he had tried to take home any 'souvenirs' of his find), the main question is not therefore whether or not this carcass existed, but what it represents. Is it a new species, or is to something more mundane? Until another of its kind is located, or the remains of the original found, a positive identification will remain elusive.

4. CONCLUSIONS

Although basilosaurus is a good candidate identity for these 'marine saurians', it is not the only archaeocete that could be conceivably extant. The larger sightings of 'marine saurians' could be put down to basilosaurs swimming. However two sightings, which have been discussed, could also possibly be explained with reference to another primitive archaeocete, namely the *Rodhocetus*. Size and appearance-wise, the 'Gambo' carcass is similar to reconstructions of *Rodhocetus* (it was thought to be the size of a large walrus). Some reconstructions even posit that it possessed orca like bi-colouration, which the 'Gambo' carcass seemed to exhibit. One inconsistency is the placement of the nostrils, as although the nostrils of *Rodhocetus* had not advanced to the top of its head to form a blowhole as in modern cetacea, they were not located on the tip of its head either, as in 'Gambo'. If anything they could be described as having occupied a position on its head intermediate of both of the aforementioned. The inconsistency could be reconciled in one of two ways; either Burnham misreported the position of the nostrils in his original specimen-based description, or 'Gambo' represents an evolutionary modification on the original *Rodhocetus* body plan. A case could be made for the latter, as terminally located nostrils could theoretically allow the animal to obtain air whilst remaining completely submerged; thus they would aid in the avoidance of large sharks, and other predators. Another, albeit completely speculative, reason for this cryptid having terminally located nostrils, is the possibility that it has gone down an entirely unprecedented evolutionary road, and in doing so has evolved a keen sense of smell, which may substitute for echolocation in its case.

The other seemingly non-basilosaur but possibly archaeocete sighting is that which was made by the crew of the *U 28*. The creature was described as possessing both webbed feet *and* a long, tapering tail. This creature also seems to conform to the basic appearance of *Rodhocetus*, although the reported size of the animal puts it in the range of a basilosaur. One of three possibilities presents itself.

Firstly this could be indicative of an interesting, evolutionarily novel move towards gigantism on the part of a cetacean species, which has retained the basic body plan of *Rodhocetus*. Secondly, the creature observed by the crew of the *U 28* may in fact have been a basilosaur, but the non-crocodile traits were simply downplayed in the minds of those that witnessed it, whilst the crocodile-like traits were correspondingly exaggerated. The third alternative is that the crew of the *U 28* did in fact observe a *Rodhocetus*, but the proportions were simply exaggerated (as was the estimate of the depth at which the *Iberian* exploded). The third alternative would seem to be the most plausible in the absence of anything other than eyewitness testimony.

Perhaps the most interesting feature of 'surviving archaeocete' based identity theories for sea serpents is their favoured status not just amongst cryptozoologists, but also amongst some zoologists, like Roy P. Mackal. It seems that there exists an interesting and *rare* broad consensus within the zoological community as a *whole* that new cetaceans are the most likely large animals to be discovered. It may not be too long therefore before the zoological community gets its first glimpse of a living archaeocete species, and the mystery of the marine saurians is laid to rest.

Super-eels: a many faceted enigma.

1. INTRODUCTION

1.1 Overview

The idea of large eel-like creatures being responsible for sea serpent sightings is an old one. The first person to try and scientifically classify these sea serpents was the early nineteenth century naturalist Constantine Samuel Rafinesque-Schmaltz who developed a classification model that included four distinct types of serpent, each of which could be described as being generically eel-like in appearance (see the introductory chapter).

Within Heuvelmans' classification model, the super-eels are probably the most generic of all of his categories. Instead of proposing a single identity theory to unite the sightings placed in this category, Heuvelmans instead suggests that 'super-eels' are in fact representative of an array of mysterious but mostly *known* creatures ranging from giant eels, synbranchids and even sharks.

This author is inclined to agree with Heuvelmans in that no single type of organism would seem to satisfactorily explain every sighting, and that - more importantly - super-eels are probably something known to science, albeit something that can on occasion attain unheard of lengths. On that basis then, this section will serve to simply illustrate the two most plausible identities for the super-eels; namely that they are caused by sightings of giant frilled sharks and oarfish.

1.2 Description

The super-eel is described as having a cosmopolitan distribution. Heuvelmans did not manage to collect much in the way of sightings data for these organisms, with only 12 of the 23 total sightings being of high quality. Because of the likely multiple identities of the super-eel, only a general description can be presented. Size-wise, Heuvelmans suggests that the curve of lengths for these creatures is bimodal, with one peak at around 30 feet, and the other at 100 feet. Although Heuvelmans is sceptical of the accuracy of length estimates at the 100 feet end of the distribution he does suggest that this could be evidence of two separate creatures with radically different body lengths.

Aside from length, this sea serpent type exhibits other characteristics that are highly variable. These include the mouth, which some witnesses report as being located terminally, whereas others report as being located ventrally, and the head, which in some reports is described as being blunt, whilst in others is described as being pointed. Another inconsistent characteristic are the pectoral fins, which are present in some sightings, but not present in others. This, Heuvelmans speculates, may be due to the fact that in some sightings where they are not reported, they are in fact present but merely being held close to the body so as to render them invisible. Alternatively they may be absent all together in some of the species of super-eel that are being seen. A body that is long and thin creates the illusion of a longish neck.

In general these sea serpents are described as being blue, brown or reddish coloured on their dorsal surfaces, whilst being white on their ventral surfaces. Heuvelmans describes an interesting regional variation in colouration in the form of speckling, which he suggests is endemic to the Mediterranean, and may constitute a form of camouflage. The skin has been compared in appearance to tanned leather, supported with dermal scales giving the appearance of segments. A translucent membranous continuous dorsal fin is often described as running the length of the body, and encompassing the tail, whilst continuing along the ventral surface.

One fairly consistent characteristic is the eyes, which are usually described as being large.

Whilst at the surface, super-eels are described as assuming a coiled posture, with the occasional undulation as it is presupposed that super-eels sighted on the surface are dying. Some reports have them fighting with sperm whales, which presumably prey upon them, where they are alleged to wrap themselves around the body of the whale in a 'strangulatory' manner and 'beat' it with their tails.

2. IDENTITY THEORIES

2.1 Overview

The two most prevalent identity theories for the super-eel are - as has been mentioned - the idea that they are variously frilled sharks, or oarfish; ironically neither of which are true eels. There are other proposed identity theories. For example Heuvelmans suggested a synbranchid identity - however, swamp-eels are intolerant of salt water, and only one species, the marbled swamp eel, is known to attain any great size (1.5 metres), which is considerably shy of the lengths reported for even the smallest super-eels. It is on this basis that a synbranchid identity can be ruled out. The species that are most likely to be behind the super-eel sightings are therefore probably marine in nature.

2.1.1 Frilled sharks

The frilled shark (*Chlamydoselachus anguineus*) is a rarely ever seen primitive elasmobranch. Their habitat is deep water (between 50 and 1,500 metres) and their distribution is cosmopolitan.

Their prey includes squid, other sharks, and deep-sea fish, which they catch with the aid of a mouth full of small needle sharp tricuspid teeth. Length-wise, the shark is capable of growing up to two metres. However, the full size range is unknown owing to the paucity of specimens available for comparison. The shark is very rarely seen near the surface, and when it is, it is usually in the process of dying. Witnesses unfamiliar with the fact that this creature is a shark, often erroneously describe it as an eel.

For frilled sharks to be a reasonable identity for the super-eel, it is necessary to suppose that the shark can attain lengths considerably in excess of the maximum which has so far been recorded. Given the fact that very little is actually known about the frilled shark other than the fact that it exists, it is possible that specimens attaining great length, exist living at considerable depths. Great length in sharks is not unknown; the megalodon is thought to have been able to reach lengths of in excess of 50 feet, making a 20-30 foot specimen of frilled shark seem all the more plausible by comparison.

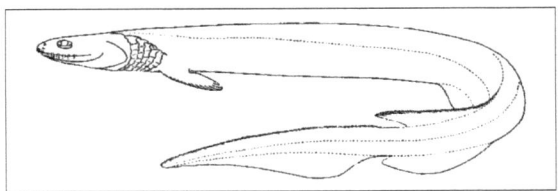

Figure 1: The frilled shark.

2.1.2 Oarfish

The oarfish, despite appearances are not actually eels. They are in fact members of the Regalecidae, a family of bony fish in the order Lampriformes. They are the largest known bony fish in the world, and it is probably no exaggeration to say that these fish are *true* sea serpents, as they have been known to reach 30 feet in length, which puts them unambiguously in the lower size range for the super-eel. They are thought to have a cosmopolitan distribution, although sightings data are rare, and much of the distribution data come from beached carcasses. They seem to live at considerable depth, between 20 to 1,000 metres, and they seem to be essentially solitary in nature, preying primarily on zooplankton, and occasionally supplementing their diets with jellyfish, squid, and small fish. It is likely that they are in turn, preyed upon by sharks and whales.

Many mysteries surround the oarfish. One report - for example - suggests that it gives off mild electric shocks when touched, suggesting an electrogenic capability. Perhaps this aids in deterring predators, or perhaps the fish employs this as part of a hitherto unknown electroreception mechanism? Strange behaviours have also been reported in this fish. For example, in 2001, an oarfish was filmed alive - and *in situ* - by a group of US Navy personnel in the Bahamas. It was utilizing an *amiiform* mode of propulsion (maintaining a rigid, vertical body whilst rhythmically beating its dorsal fin). It is thought that this behaviour is in fact a feeding posture that the fish assumes, so that it can more easily detect prey by taking advantage of sunlight entering the water. Size is also another issue of controversy with these animals, as although the largest specimens caught were in the range of 30 feet, it is possible that they are capable of growing to considerably greater size. Perhaps it is even the case that somewhere in the oceans there exist specimens of this fish that have attained sizes of 100 feet or more.

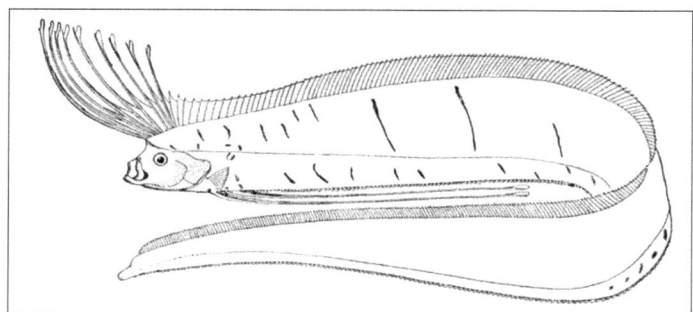

Figure 2: *The oarfish.*

2.2 Comparisons

On balance the oarfish are far more likely to be involved in sightings of super-eels than abnormally large frilled sharks, as there is a precedent for oarfish being able to attain the proportions described in encounters with this sea serpent category. However, on a basic level, it may be possible to use the described variation in head shape as a means of determining which type of animal is involved in which encounter. For example, oarfish have more rounded heads with large eyes, whereas the frilled shark has a more pointed head. As has been mentioned, both facial structures are described in encounters with super-eels.

It is also possible that other creatures are involved in super-eel sightings, giant morays and conger eels, for example, may be candidate species, as they are both known to be capable of growing up to several metres in length. One argument *against* the possibility that true eels are the culprits behind super-eel sightings is actually based on a misconception held by Heuvelmans and others during the 60's about allometric scaling in juvenile eel leptocephalus vs. adult forms.

In 1930 the ship *Dana* trawled up a 6 feet long specimen of leptocephalus (eel larva) between the Cape of Good Hope and Saint Helena. One of the scientists on board, Anton Bruun estimated that the adult form could reach sizes of 108 to 180 feet. Heuvelmans promotes this as proof positive of giant eels. However, what neither Bruun nor Heuvelmans could have known, is that the allometric scaling ratio between juvenile leptocephalus and adult eel, seems to break down over a certain size, and that whereas small eel leptocephali can attain many times their size in their adult forms, large leptocephalus tend to be only slightly larger when they reach maturity.

3. A CURIOUS SIGHTING

3.1 Overview

Included here is a most interesting sighting of a super-eel seemingly engaged in a most curious behaviour - fighting with a sperm whale. This atypical account has to be considered carefully in light of the above-mentioned identity theories, as it is highly incompatible with what is known about the behaviour of both frilled sharks and oarfish.

3.2 The sighting

Year: 1875.
Location: Cape São Roque.
Number of witnesses: Numerous (unknown).

On the 8th of July 1875, the entire crew of the cargo ship *Pauline*, were witness to what seemed to be a titanic struggle between a sperm whale and a sea serpent. Five days after witnessing the battle, an identical serpent was spotted in the waters 200 yards from the ship. After the ship arrived in Zanzibar, the Rev. D. L. Penny, compiled a report of what was based upon witness statements and produced a sketch, all of which was sent to the *Illustrated London News* for publication. On pages 262 and 263 of *In the Wake of the Sea-Serpents*, Heuvelmans recounts the much more detailed and interesting first hand account of Captain Drevar, which was made for the *Pauline's* log book.

"The weather fine and clear, wind and sea moderate. Observed some black spots on the water, and a whitish pillar, about thirty feet high above them. At the first glance I took all to be breakers as the sea was splashing up fountain-like about them, and the pillar a pinnacle rock, bleached with the sun; but the pillar fell with a splash, and a similar one rose. They rose and fell alternately in quick succession, and good glasses showed me it was a monstrous sea-serpent coiled twice round a large sperm whale. The head and tail parts, each about thirty feet long, were acting as levers, twisting itself and victim round with great velocity. They sank out of sight about every two minutes, coming to the surface still revolving; and the struggles of the whale and two other whales, that were near, frantic with excitement, made the sea in their vicinity like a boiling cauldron; and a loud and confused noise was distinctly heard.

This strange occurrence lasted some fifteen minutes, and finished with the tail portion of the whale being elevated straight in the air, then waving backwards and forwards, and lashing the water furiously in the last death struggle, when the body disappeared from our view, going down head foremost to the bottom, where no doubt it was gorged at the serpent's leisure; and that monster of monsters may have been many months in a state of coma, digesting the huge mouthful. Then two of the largest sperm-whales that I have ever seen moved slowly thence towards the vessel, their bodies more than usually elevated out of water, and not spouting or making the least noise, but seeming quite paralysed with fear; indeed, a cold shiver went through my own frame on beholding the last agonizing struggle of the poor whale that had seemed as helpless in the coils of the vicious monster as a small bird in the talons of a hawk. Allowing for two coils round the whale, I think the serpent was about 160 or 170 feet long, and 7 or 8 feet in girth. It was in colour much like a conger-eel; and the head, from the mouth being always open, appeared the largest part of its body."

Figure 3: *The Pauline encounter as visualized by the Rev. D. L. Penny.*

On the second occasion of its sighting Drevar feared for the safety of the ship after having witnessed what the serpent had done to the sperm whale.

"As I was not sure that it was only our free board it was viewing, we had all our axes ready, and were fully determined, should the brute embrace the Pauline, *to chop away for its backbone with all our might, and the wretch might have found for once in its life that it had caught a Tartar."*

3.3 Considerations

The Drevar account, as has been mentioned, would seem to indicate behaviour that is atypical for both oarfish and frilled sharks. It seems to suggest that something different may lay behind this particular sighting; perhaps something with the characteristics of a conger-eel. One of the animals that springs to mind when thinking of sperm whales is *Architeuthis*, and one interpretation of this encounter is that the thing coiled around the sperm whale is not the body of some colossal super-eel, but the tentacle of a very large squid. However, as Heuvelmans points out, the sighting possesses two characteristics that are simply inconsistent with a giant or colossal squid identity. Firstly, the creature's mouth was described as always being open, which is visually inconsistent with the theory that the head of the serpent could have been the 'swelling' at the tip of a tentacle.

Also, Penny describes the colouration as being darkish brown on top with a white belly - a scheme that is inconsistent with the known colouration of squid tentacles. An alternative explanation is that Drevar and his crew, were witnessing events correctly, but got the cause and effect confused.

What if the sperm whale, instead of being attacked by the super-eel, was in fact attacking it?

Perhaps the whale accidentally became entangled whilst trying to feed and eventually managed to wriggle free of its abortive meal which was then seen by the crew of the *Pauline* five days later injured, and dying at the surface.

4. CONCLUSIONS

The super-eel seems to be generally enigmatic, exhibiting an assortment of traits that on the one hand seem to allow its identity to be pinned down to known organisms, yet on the other hand seem to suggest something else entirely. The identities of the various species that make up this category of sea serpent seem as illusive as the super-eel itself.

Much work needs to be done on developing a viable taxonomy for the super-eel category. Sightings of known organisms, like oarfish, need to be sifted out from sightings that are clearly inconsistent with such an identity. Initially there could be two categories - *oarfish* and *giant forms;* the latter can be a dustbin category for creatures that appear to be large versions of frilled sharks, conger and moray eels etc. Through a more precise correlation of characteristics, a more detailed taxonomy can be built up for the members of this *giant forms* category, and a more detailed picture can emerge of the true diversity of super-eel forms.

Giant invertebrates: The most plausible category.

1. INTRODUCTION

1.1 Overview

The category of 'giant' or 'elongated' invertebrates was a relatively recent addition to Heuvelmans' classification scheme. It was devised only after the publication of *In the Wake of the Sea-Serpents,* and represented a considerably more 'sober' departure from the more speculative and 'outlandish' categories presented previously. Despite the seemingly general nature of the term 'invertebrate', it has a somewhat more specific meaning in the context of this category.

Invertebrate here refers largely to elongated and oversized examples of the sub-phylum Urochordata (which includes organisms like salp) and the phylum Ctenophores (sea-gooseberries, Venus' girdle etc.). The giant invertebrates could theoretically be expanded to accommodate anomalously large examples of the phylum Cnidaria (jellyfish and their allies) and the phylum Nemertea (ribbon worms), without the category vying too far from Heuvelmans' original intentions.

1.2 Description

Cryptids placed in this category can generally be described as anomalously large examples of known organismal forms, which can be collectively, characterized as 'primitive' invertebrates. The cryptids in this category may come in a multitude of shapes and sizes, and may include things like vast strings of salp colonies that may be hundreds of feet in length far surpassing anything observed to date. Or giant examples of things like Venus' girdle, a ctenophore, or even examples of jellyfish such as the Arctic lion's mane and bootlace worms exceeding maximum-recorded dimensions. Heuvelmans was not explicit about whether or not the giant invertebrate category was meant to contain only new species, or simply exaggerated versions of existing species, making the category somewhat vague. It is probably safe to assume however that this category was meant for both, and that Heuvelmans intended it as an 'embryonic' or 'dustbin' category into which sightings could be lumped with a view to future disambiguation and possible reclassification. Unfortunately, however, there has been no further work in this direction.

2. IDENTITY THEORIES

2.1 Disentangling the giant invertebrates

In this section, representative species of the four proposed phyla comprising the giant invertebrate category, will be described with the intention of building a case for the existence of abnormally large variants.

2.1.1 Urochordata – salp

Salp, are free floating tunicates whose individual bodies are shaped somewhat like a barrel. In terms of their developmental biology, structural organization and genetics they are more closely related to vertebrates than to jelly fish, to which they bear a cursory resemblance. Salp contract themselves, forcing water through their bodies which facilitates movement, and allows them to filter out phytoplankton, which is their primary food source. They are very fast growing, and reproduce via cloning, becoming highly abundant during periods in which phytoplankton populations are blooming. Salps can form long string-like colonies, which can be many tens of metres in length in some instances.

Figure 1: A salp string.

2.1.2 Ctenophora – Venus' girdle

The phylum Ctenophora contains around 100 species, one of the most remarkable of which is the Venus' girdle (*Cestum veneris*). This organism is a ribbon-shaped comb jelly found in the Mediterranean. Its body is of a distinctive violet colouration and it can grow to about a metre and a half in length. It possesses a muscular system, which it uses to propel itself in an undulating manner. In nearly every respect it is abnormal for a ctenophore; it is much larger than any other species in this phylum, also its ribbon-like body plan is unique.

Venus's girdles feed on plankton, free swimming larvae, other Ctenophores, and even small fish. These are captured with the aid of colloblasts, which disgorge sticky threads when stimulated by contact with prey.

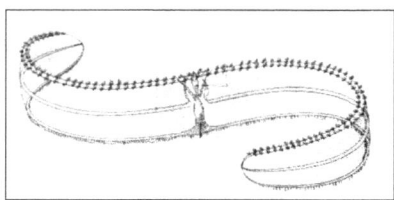

Figure 2: Venus' girdle.

2.1.3 Cnidaria – The Arctic lion's mane jellyfish

Confined to the cold northern waters of the Pacific, Atlantic, and Arctic oceans, the lion's mane jellyfish (*Cyanea capillata*) already holds the record of being one of the longest animals in the world, with the current record holder being a specimen discovered in 1870 which measured 120 feet in length with a bell diameter of around seven feet. They are pelagic for the most part, drifting with the oceanic currents and feeding on small fish, ctenophores, and zooplankton. Turtles, sea birds, and large fish, in turn prey upon these jellyfish. One point of interest is that these jellyfish provide an ecological oasis for a variety of crustacean and fish species, which live in - and amongst - the tentacles, taking advantage of the protection afforded them, and the regular supply of food.

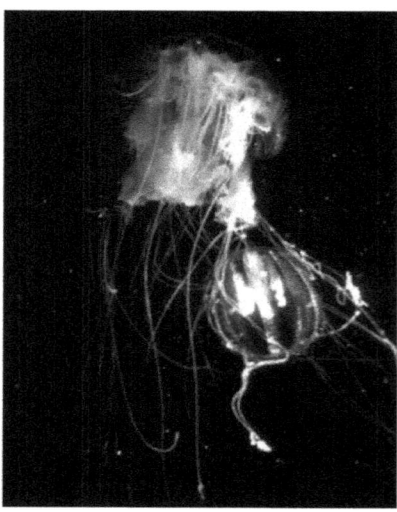

Figure 3: Lion's mane jellyfish feeding on a sea gooseberry.

2.1.4 Nemertea – The bootlace worm

The phylum Nemertea contains around 1,400 species of mostly marine-dwelling worms, with a few freshwater, and a small number of terrestrial, specimens also. The bootlace worm (*Lineus longissimus*) is a particularly remarkable species, as it has been known to reach lengths of up to 180 feet, as in the case of a storm-beached specimen recovered in 1864 from a beach in St. Andrews, Scotland, making it the longest known animal in the world. Like many ribbon worms, the bootlace worm feeds with the aid of an extendable proboscis equipped with a cluster of sticky filaments, which are used to ensnare prey.

Like many ribbon worms, the bootlace worm feeds with the aid of an extendable proboscis equipped with a cluster of sticky filaments, which are used to ensnare prey.

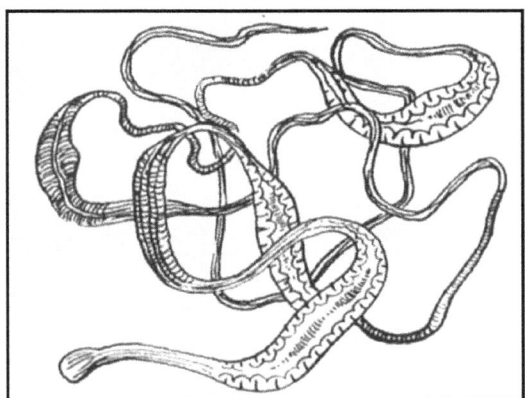

Figure 4: Bootlace worm after Heuvelmans (1968).

2.2 Discussion

In order to establish the likelihood of there existing abnormally large variants of species in the above listed phyla it is necessary to look at a variety of factors regarding the life history and ecology of these creatures.

Length is a function of a variety of different factors in each of the aforementioned organisms. In salp colonial strings, 'length' is not a property of a single organism, but is instead a property of the additive effect of individual salp linking to form a colonial string. Therefore the length of the strings is not subject to restrictions on size in the member organisms, but on the various ecological stresses that encourage recruitment into strings, such as nutritional scarcity. The maximum length of a hypothetical string of salp is also subject to mechanical factors, such as points of linkage weakness that could facilitate the break up of salp strings subjected to high mechanical stress environments, and changes in ecological stress patterns that could cause disaggregation.

It is difficult to predict the theoretical maximum length of a salp string, but assuming a high degree of environmental homogeneity, it may be the case that they could reach many hundreds of feet in length.

The other three candidate identities are all organisms that must attain great length *as* organisms rather than as part of a colony in order to qualify for the giant invertebrate category; so are therefore subject to different constraints and factors.

For most invertebrates there exists no size curve, as in the case of mammals, where maximum length is largely determined by developmental genetic constraints. Many invertebrates simply follow a linear pattern of size increase over time, where the constraints are primarily in the form of environmental effects such as nutritional availability and factors influencing lifespan such as disease and predation. Another factor affecting the size of invertebrates is the degree to which reproductive maturity is delayed, this being the primary cause of deep-sea gigantism.

There is little debate in zoological circles that lion's mane jelly fish in excess of 120 feet and bootlace worms in excess of 180 feet probably exist, [16] however it would be harder to make a case for the existence of massively oversized examples of Venus' girdle, as their maximum observed size seems to be in the region of 1.5 meters. Although it is of course possible that there could exist a Venus' girdle-like species of Ctenophore, which has attained considerably greater size and is occasionally seen at the surface. The case for this is made all the more compelling by the fact that the species diversity in this phylum is likely much higher than currently thought. This new ctenophore species may be deep sea dwelling, and may also be responsible for a small number of super-eel sightings due to its flattened body, purple colouration and the undulating manner in which it moves.

---

16   Although it *must* be noted that the bootlace worm can increase its length considerably through stretching, so reports of extreme length have to take this into consideration.

3. CONCLUSIONS

On pages 221 and 222 of *In the Wake of the Sea-Serpents* Heuvelmans recounts an interesting incident involving some fishermen in 1849, that bought what they thought was a young sea serpent to the secretary of a local museum. *The Montrose Standard* covered the event:

"The animal, whatever it may be called, is still alive, and we have just been favoured with a sight of it; but whether it really be a young sea-serpent or not, we shall leave those who are better acquainted with Zoology then we are to determine. Be it what it may, it is a living creature, more than 20 feet in length, less than an inch in circumference, and of a dark brown chocolate colour. When at rest its body is round; but when it is handled it contracts upon itself, and assumes a flattish form. When not disturbed its motions are slow; but when taken out of the water and extended, it contracts like what a long cord of caoutchouc [17] would do, and folds itself up in a spiral form, and soon begins to secrete a whitish mucous from the skin, which cements the folds together, as for the purpose of binding the creature into the least possible dimensions." As Heuvelmans observes, this particular specimen was dismissed as a bootlace worm long ago by Edward Newman, however this tale served to illustrate Heuvelmans' belief that: *"in cryptozoological research, whenever a sea-monster can possibly be explained as a known animal it should be immediately eliminated."* Based on this it appears that Heuvelmans may have in fact intended the creatures in his giant invertebrates category to be new species only, however it is also seems to be the case that at some point between the publication of *In the Wake of the Sea-Serpents* and *The Kraken and the Colossal Octopus: In the Wake of Sea-Monsters*, Heuvelmans changed his mind about this statement and instead decided to include abnormally large forms of known animals as cryptids. [18]

It is unfortunate that Heuvelmans had only the opportunity to write very little about speculative giant invertebrates, as these do seem to be the most plausible of all of the sea serpent categories in terms of actual specimens and what is known of the biology of these invertebrates. As has been mentioned, up until this writing, there has been no follow up work on giant invertebrates. The category or one like it is missing from both the Coleman & Huyghe and Champagne classification schemes, which in every other respect follow Heuvelmans classification system quite closely.

It is possible that giant invertebrates of the type discussed here have simply attracted little in the way of attention as they are certainly not as glamorous as the long necked seals, or the marine saurians for example, yet they *are* highly plausible as a category, and giant invertebrates currently *unknown* to science have probably been the source of a considerable number of sea serpent sightings in the past through to the present day, so warrant inclusion on that basis.

---

17 Natural rubber.
18 This is admittedly somewhat of a paradox as Heuvelmans entertained the possibility of known creatures being at least partly responsible for sightings of super-eels also. Perhaps to clarify further, Heuvelmans may have initially reserved this disclaimer for normally-sized, *known* species only, and allows for abnormally sized known species to count as cryptids – eventually warming to the idea sufficiently to introduce a separate category for them in the form of the giant invertebrates.

End notes.

1. SUMMARY

1.1 Overview

The end verdict of this reassessment of Heuvelmans' sea serpent identity theories, is simply that some of the proposed identities seem to adequately correspond to something plausible (in other words Heuvemlmans probably got it right the first time), whilst others should be re-interpreted as something else. Obviously the knowledge of hindsight has much to do with this. Armoured archaeocetes and super-sized eels coming from 6-feet long leptocephalus larvae were considered reasonable in the 60's and 70's, for example, whereas today we know better.

The additional factor that has allowed these determinations to be made is the use of the plausibility method. For some of the cryptids, there seem to exist some fascinating ecological correlates which emerge from comparing the sightings data with data from other sources (such as cod stock data in chapter one) also, overlooked but seemingly solid evolutionary narratives have presented themselves in some cases, allowing for realistic scenarios to be posed explaining the origins of certain cryptid types (for example linking the *con rit* with Arthropleura).

These extrinsic factors feed back into the identity theories be they old or new, grounding them and making them reasonable in a way that compliments the available data.

1.2 Comparative table of identities

Name (Heuvelmans).	Identity (Heuvelmans).	Name (Woodley).	Identity (Woodley).
Long necked seal (*Megalotaria longicollis*).	Pinniped, otariid.	Retained.	Retained.
Merhorse (*Halshippus olai-magni*).	Pinniped, unknown family.	Retained.	Pinniped, otariid.
Many humped sea serpent (*Plurigibbosus novae-angliae*).	Archaeocete.	Retained.	Reassigned to the lutrinae.
Super-otter (*Hyperhydra egedei*).	Primitive archaeocete or sirenian.	Retained.	Reassigned to the lutrinae.
Many finned sea serpent (*Cetioscolpenda aelani*).	Armoured archaeocete.	Con rit, suggested reclassification as *Mariascolpendra aelani*.	Archaeocete identity dropped; replaced with invertebrate myriapod identity in the class Arthropleuridea.
Super-eel.	Possible multiple eel, synbranchid and elasmobranch identities.	Retained.	Elasmobranch and eel identities retained, synbranchid identity dropped; oarfish identity added.
Marine saurian.	Thalattosuchian crocodile and mosasaurus.	Primitive cetacean.	Reptile identities dropped; replaced with basilosaurus and rodhocetus.
Giant invertebrates.	Salp colonies and Venus' girdle.	Retained.	Salp colonies and Venus' girdle retained; giant Cnidaria and Nemertia identities added.

1.3 'Dropped' categories

This author has 'dropped' the 'yellow belly' and 'father-of-all-the-turtles' categories, primarily on the same grounds as Heuvelmans; namely that there is simply insufficient data to reasonably maintain a case for the existence of these cryptids. A strong case can also be made for rejecting them in the context of the plausibility method, as the data that exists on these cryptids is far too inconsistent to allow for a reasonable ecological niche to be determined and therefore a plausible identity to be inferred. However it is worth noting that the 'Father-of-all-the-turtles' category has enjoyed something of a revival in the form of the 'cryptic chelonians' category of the Coleman & Huyghe model, and as the 'type five or 'carapaced' sea serpent [19] in the current Champagne scheme. Perhaps some new and convincing data will come to light, which could vindicate this 'lost' category after all.

1.4 Additional categories and future research directions

It is obvious that Heuvelmans' categories are incomplete as they stand, and are therefore not truly representative of the full potential diversity of marine cryptid forms; however the development of new categories was *not* the purpose of this manuscript, which was to simply re-evaluate and 'modernize' Heuvelmans' existing categories. The former would require much more research by this author than has been done to date, and may form the basis of a subsequent publication building on this text. In terms of plausible but tentative additions to Heuvelmans' scheme however, 'mystery cetaceans' possessing double or elongated dorsal fins and 'mystery elasmobranchs' would seem to be a fair indication of the sort of form that hypothetical additions could take, as perhaps best exemplified by their inclusion in the classification scheme of Coleman & Huyghe.

1.5 Fossil evidence and ghost lineages.

One of the main objections commonly raised against the kind of speculative arguments articulated in this monograph, and in all other attempts to address the same issue of classification, is the seeming absence of fossil evidence suggesting the evolution of these speculative creatures. The only possible exception to this is in the form of the swan necked seal (see figure twelve in chapter one), which although probably not on a line of descent to the long-necks, demonstrates convincingly that pinnipeds have - at some point in their evolution - moved towards acquiring plesiosaur-like long necks as an aid to hunting. When regarding ghost lineages, it is worth considering that the chances of any organism getting fossilized are astronomically slim, and it is certainly the case that for every species which gets represented in the fossil record many more do not. Due to this, whole evolutionary lineages are lost to time, which means that the absence of a ghost lineage cannot be considered a definitive argument against the existence of a particular cryptid.

---

19 Where it is described as an armored mammal, rather than a reptile.

2. AFTERWORD

This manuscript was in fact originally intended to be 'many' manuscripts. I had written the first chapter on Heuvelmans' pinnipeds with the naïve hope of getting it published as a small book, but this expectation was dashed by the fact that although I received good feedback from a variety of sources (including Chad Arment) it was deemed too short to publish as a stand-alone text.

I then got in touch with Jonathan Downes of the CFZ, and after impressing him with big, sciencey words like 'competitive exclusion' managed to convince him to take a look at my manuscript. He was duly impressed, and offered to give it pride of place in the 2008 yearbook and - for a while - that's how things stayed. But one day I got the writer's bug, and decided to put together a shorter manuscript on a more data-poor cryptid, namely the enigmatic *con rit*. I then sent this to Jonathan with a view to getting it included in *Animals and Men*. After ringing him up one day to moan about not having heard back from him, or something or other, we got round to discussing how the ecology-themed methodology developed for the pinnipeds and *con rit* articles could be developed into a more coherent methodology for assessing lesser known cryptids in general.

At some point in the conversation, one of us (I don't remember which one) came up with two truly catastrophic ideas; firstly to write a 'pilot' paper in which the 'plausibility method' (as it became known) could be used to infer an identity for a lesser known cryptid that frankly no one really had a clue how to approach - such as, say, the Mongolian deathworm - and secondly, applying this to *all* of Heuvelmans' cryptid categories with a view to the creation of a book! So, after generating a ton of research, consuming an equivalent mass of caffeine, and getting more peculiar looks from my PhD supervisor than can possibly be good for me, presto, here it is. I enjoyed every minute of it!

Appendix

Provided here are more detailed descriptions of the various extinct taxa mentioned in the text.

Ambulocetus

Ambulocetus is an extinct genus of primitive, amphibious cetacean containing one known species, *Ambulocetus natans*, which was first described by Anatomist and Palaeontologist Johannes Thewissen and colleagues in 1994. It lived during the Eocene epoch approximately 49 million years ago in what is now Pakistan, and bore a striking resemblance to the modern crocodiles, both morphologically and ecologically. As an ambush predator of moderate length (3 metres) it would have exploited shallow bodies of water for hunting; the absence of external ears means that it must have been able to detect underwater vibrations through the periotic bones of its inner ear in order to locate prey.

Data obtained from a chemical analysis of its dentition indicate that it had the ability to transition between salt and freshwater environments, suggesting a capacity for osmoregulation.

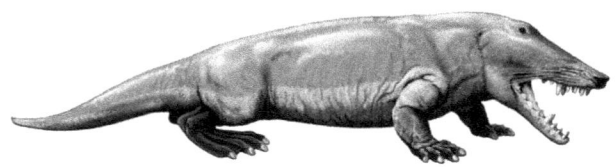

Arthropleura

The Arthropleuridea are a possibly extinct class of terrestrial and aquatic arthropods, which based on phylogenetic research, are thought to be representatives of the sub-phylum Myriapoda. The class is thought to be descended from a crustacean-like ancestor. Previously the Arthropleuridea had been aligned with the sub-phylum Chelicerata, which contains the arachnids and horseshoe crabs. The most well known representative of the class is the genus *Arthropleura*, which contained the largest species of terrestrial arthropods ever to have existed.

The arthropleura lived during the Upper Carboniferous between 340-280 million years ago, attaining sizes of up to 3 metres in length. Their primary habitat was dense forest, however they are thought to have been amphibious also, using bodies of water to moult.

Their diet is thought to have been primarily herbivorous based on the identification of pollen spores observed in the guts of fossil specimens, however it is possible that the larger species were omnivorous also.

The large size of arthropleura has been attributed to the absence of terrestrial predators and the presence of higher amounts of atmospheric oxygen. Respiration may have involved the use of gills or lungs, as spiracles are absent from fossil remains.

It is thought that arthropleura succumbed to a climate change event in the early Permian period, which was characterized by increasing aridity and habitat loss.

Elasmosaurs.

The Elasmosauridae are an extinct family in the extinct order Plesiosauria, which existed up until the end of the Cretaceous period. The most well known example of the Elasmosauridae is *Elasmosaurus platyurus*, one of eight known species in the genus *Elasmosaurus*, which was discovered in 1868 by the legendary Palaeontologist Edward Drinker Cope in Kansas. It was notable for having the longest body length of any known plesiosaur species, reaching lengths of up to 14 metres (46 feet), and weighing in excess of two tons. Its neck, which accounted for fully half of its overall length, contained in excess of 70 vertebrae arranged into a long, brittle column supporting a small head containing sharp, outward pointing teeth, which overlapped when its jaws were closed. Its primary food sources appear to have been bony fish such as *Enchodus* and molluscs such as ammonites and belemnites.

For a long time it was believed that Elasmosaurs came ashore to lay eggs in a manner similar to turtles, however recent fossil finds indicate that they were viviparous; giving birth to live young at sea.

Enaliarctos

Enaliarctos is an extinct genus containing five known species of pinniped that lived in the late Oligocene epoch approximately 22 million years ago. The species in this genus comprise the oldest known pinnipeds. Enaliarctos is especially significant as the consensus view in Palaeontology is that it may in fact represent the common ancestor of all three known pinniped families, which potentially puts it on a direct line of ancestry to modern pinnipeds.

Enaliarctos combines characteristics from all three known pinniped families. It used both its fore limbs and hind limbs for aquatic locomotion, which represents the ancestral condition for pinnipeds. Members of the Phocidae family use only their hind limbs, whereas members of the Otariidae use only their fore limbs. Members of the Odobenidae use both sets of limbs. Other modern pinniped innovations present in Enaliarctos include a sophisticated inner ear design that would have enhanced their hearing underwater and large eyes, both of which are found in the otariids.

The presence of slicing carnassials indicate that once it caught an item of prey it probably had to return to land in order to feed. One species in this genus, *Enaliarctos emlongi,* is notable for having been named after the fossil hunter, Douglas Emlong.

Enchodus

Enchodus is an extinct genus of bony fish from the upper Cretaceous. Edward Drinker Cope discovered the species *Enchodus petrosus*, in 1874.

One of the distinguishing morphological features of this genus is the presence of long, fang-like teeth located on the palatine bones, indicating that enchodus were predatory. Enchodus species range in size from being nearly 2 metres in length to as little as only a few centimetres and remains of the fish have been found in the stomach contents of a variety of large marine predators such as elasmosaurs and mosasaurs.

Ichthyosauria

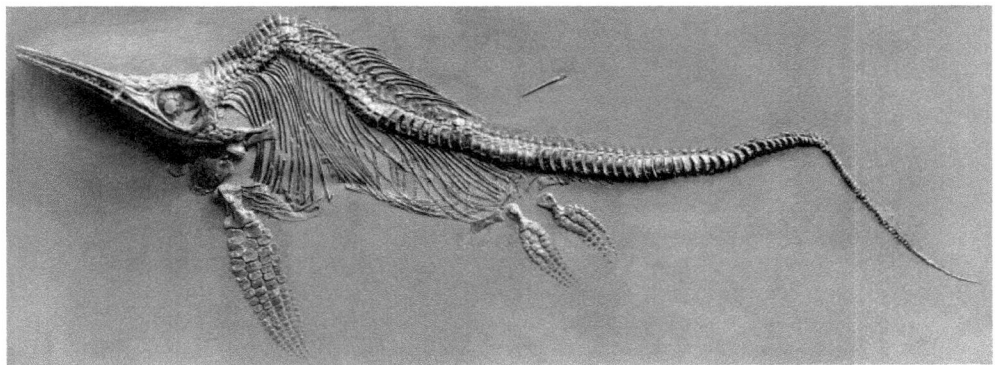

The order Icthyosauria is comprised of a phylogenetically and morphologically diverse range of marine reptiles that lived in the Mesozoic era. They first appeared in the Middle Triassic, 230 million years ago, and lasted until the Cretaceous, 90 million years ago, when their diversity was whittled down to a single genus.

Their peak abundance was during the Jurassic, where the order exhibited its largest diversity of morphological forms. Although the average length of an ichthyosaur was between two and four metres, several species grew to considerably greater sizes.

The Ichthyosauria are most remarkable for being highly convergent upon the niches currently occupied by members of the order Cetacea. Some examples, such as the ichthyosaurs in the Cretaceous genus *Platypterigius*, are extremely close morphologically to the modern dolphins, as they possess highly streamlined, smooth bodies with a powerful tail equipped with a vertical shark-like fin for propulsion at considerable speeds (up to 25 miles per hour). Like the plesiosaurs and the cetacea, ichthyosaurs gave birth to live young at sea.

One significant way in which the ichthyosaurs do not seem convergent with the cetacea is in the evolution of the melon organ, which gives many species of dolphin and toothed whale the ability to echolocate. Many ichthyosaurs did however possess large eyes situated within bony orbits to protect against extreme pressure, which indicates how another sense can be co-opted and enhanced for the same purpose – namely hunting.
One of the factors that led to the extinction of this order was the evolution of highly evasive teleost fish, which ichthyosaurs were simply unable to catch. Plesiosaurs and mosa-

saurs eventually came to replace them as the top predators.

Kutchicetus

Kutchicetus minimus is an extinct species of primitive cetacean in the Remingtonocetid family, first described in 2000 from a specimen recovered from India by Johannes Thewissen and colleagues.

The species lived during the early Eocene epoch between 46 and 43 million years ago. They are thought to have been closely related to Ambulocetus, however unlike Ambulocetus, they are believed to have been more specialized for an aquatic lifestyle and much closer in size to the modern common otters (2.5 metres).

Their tails are thought to have been flattened in a fluke-like manner so as to aid them in underwater propulsion. Like modern otters, they are thought to have exploited shallow water, ambushing prey with the help of their elongated snouts, which were equipped with sharp teeth to aid in the capture of fish.

Mosasaurs

The Mosasauridae are an extinct family of large bodied marine lepidosaur reptiles first identified from the fossil remains of a skull found in 1780 at a limestone cave in Maastricht on the Meuse. The family contains 27 genera, the most well known representative of which is the genus *Mosasaurus*, which lived during the late cretaceous period, around 70 to 65 million years ago.

The mosasaurs possessed long, sinuous bodies equipped with fore and hind limbs that had been modified into flippers. These flippers played no role in propelling them underwater, they merely aided in steering. The main propulsive force came from the mosasaurs' muscular tail, which was flattened in the vertical plane.

Mosasaurs ranged in size from just a couple of metres in length to over 15 (in the case of the genus *Mosasaurus*) and appeared to have been highly predatory, possessing a powerful crocodile-like set of jaws and strong neck muscles designed for tearing and ripping their prey apart. Their diet is thought to have been composed of turtles, ammonites and fish.

Phylogenetic investigations indicate that the mosasaurs evolved from aigialosaurs, close relatives of modern monitor lizards, whilst cladistic analyses of the structure of the mosasaurs skull and jaw indicate that it may be even more closely related to modern snakes, as there are considerable symptomatic similarities between the two.

Rodhocetus

Rodhocetus is a possibly extinct genus of walrus-sized, primitive, semi-aquatic cetaceans in the Protocetidae family that lived about 47 million years ago. The genus contains two known species; the most recently discovered of which, *Rodocetus balochistanensis* was first described from fossil remains recovered in Pakistan, by Palaeontologist Philip Gingerich in 2001.

Rodhocetus possessed a barrel shaped torso with limbs that while still suitable for land-based locomotion, had been partially modified into flippers. Its skull was long and crocodile-like, possessing a number of differentiated teeth, which indicate a predatory lifestyle. Rodhocetus possessed modern whale-like ear bones and is thought to have utilized its hind legs as its primary form of propulsion. There is debate as to whether or not its tail terminated in a fluke, but in either case it is only likely to have served as a rudder rather than as a significant source of power for propulsion.

Recovery of the ankle bones of *Rodhocetus balochistanensis* indicate an evolutionary link to the Artiodactyls (such as pigs and hippopotamuses), which challenges the long held idea that the Cetacea are descended from the carnivorous Mesonychia.

Swan necked seal

The Swan necked seal, or *Acrophoca longirostris* is an extinct species of phocid seal that lived during the Miocene and Pliocene epochs in what is today the Pacific coast of South America. The Palaeontologist Christian de Muizon first described the species in 1981. It has been allied with the Monachines, which include such phocids as the elephant seal and the monk seal.

Appearance wise this seal was atypical in that it possessed a reasonably long, flexible neck that made up 21% of its body length. This would have caused it to resemble a 'mammalian plesiosaur' of sorts. Many seals can elongate their necks so as to strike at prey, it is possible that the swan necked seal used its neck in a similar manner to the plesiosaurs, namely to infiltrate shoals of fish and strike at them.

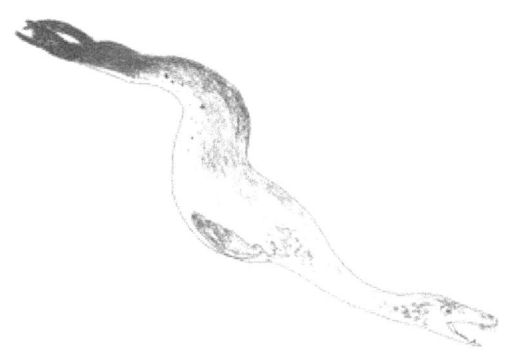

It is thought that there are additional species in the genus *Acrophoca*, although much work apparently needs to be done in describing them. One purported species, *Acrophoca piscophoca* appears to have been even more plesiosaur-like than the swan necked seal.

Thalattosuchia

Much debate surrounds the exact placement of the Thalattosuchia in modern classification schemes. Originally coined by the Palaeontologist Eberhard Fraas in 1902, the term has been used to designate a suborder or an infraorder in the now obsolete paraphyletic group Mesosuchia. These days the term is simply considered a clade, or a non-taxonomically specific alliance of two related families of large marine crocodylomorphs, the Metriorhynchidae and the Teleosauridae, which lived from the early Jurassic to the early Cretaceous.

These crocodiles varied in terms of their preferred habitats. Early examples of what are thought to be representatives of the Thalattosuchia were partly aquatic, however by the Cretaceous representative species of both families had become mostly aquatic and largely adapted to life as pelagic marine predators. In adapting to this niche, they appear to have lost their osteoderms (armour) as a trade off for greater hydrodynamic efficiency. They also appear to have evolved ichthyosaur-like tail fins, which would have helped drive them through the water.

Zeuglodon/Basilosaurs

The basilosaurs are a genus of primitive serpantiform whales containing three known species that are thought to have lived in the Eocene between 40 and 34 million years ago, when they are popularly believed to have become extinct. The first fossil basilosaurs were recovered from Louisiana, America in the early part of the 19th century. It was initially mistaken by the Anatomist Richard Harlan, for a reptile in 1834 and was christened with the genus name of *Basilosaurus,* literally meaning 'King lizard'. When the infamous British Naturalist Sir Richard Owen examined the remains he correctly proposed a mammalian identity and suggested that they be re-classified as belonging to the new genus *Zeuglodon,* which was named for their double rooted 'yoke-like' teeth, a feature common in marine mammals. Although this cetacean is referred to formally by the genus name of *Basilosaurus,* the technically more accurate name of Zeuglodon still has common currency.

The primary feature of this whale was its snake-like body, which was uniquely flexible, both in the vertical and horizontal planes, and could reach lengths of up to just over 60 feet. Their heads were also remarkable, as they resemble to a degree those of the large marine reptiles, which died out tens of millions of years before them. The basilosaurs are not thought to have had the same capacities as modern whales for diving or for endurance swimming, being much weaker and more anatomically underdeveloped in general. Their preferred habitat would have been closer to the surface where they would have hunted for fish, sharks, squid and turtles. It is still a matter of some debate as to whether or not the small, primitive hind limbs possessed by the basilosaurs were vestigial or whether they in fact functioned as guides used during copulation.

GEOLOGIC TIME SCALE

EON	ERA	PERIOD		EPOCH	Age (Ma)
Phanerozoic	Cenozoic	Quaternary		Holocene	Present
					0.01
				Pleistocene	
					1.6
		Tertiary	Neogene	Pliocene	
					5.3
				Miocene	
					23.7
			Paleogene	Oligocene	
					36.6
				Eocene	
					57.8
				Paleocene	
					66.4
	Mesozoic	Cretaceous			
					144
		Jurassic			
					206
		Triassic			
					245
	Paleozoic	Permian			
					286
		Pennsylvanian (Carboniferous)			
					320
		Mississippian			
					360
		Devonian			
					408
		Silurian			
					438
		Ordovician			
					505
		Cambrian			
					570
Precambrian	Proterozoic				
					2500
	Archean				
					3800
	Hadean				
					4550

Age in millions of years before present

Sources and further reading

Books

Aelianus, C. (?). Περί Ζώων Ιδιότητος (*De Natura Animalium*). Complete English translation available at http://penelope.uchicago.edu/Thayer/E/Roman/Texts/Aelian/home.html retrieved on 13/01/08.
Arment, C. (2004). *Cryptozoology: Science and Speculation.* Coachwhip Publications, PA, USA.
Bord, J., and Bord, C. (1987). *Modern Mysteries of Britain.* Guild Publishing, London, UK.
Coleman, L., and Huyghe, P. (2003). *The Field Guide to Lake Monsters, Sea Serpents and Other Mystery Denizens of the Deep.* Tarcher, NY, USA.
Costello, P. (1974). *In Search of Lake Monsters.* Garnstone Press, London, UK.
Ellis, R. (2006). *Monsters of the Sea.* (1st ed). The Lyons Press, CT, USA.
Gould, S. J. (2002). *The Structure of Evolutionary Theory.* The Belknap Press of Harvard University Press. London, UK.
Gudger, E. W. (1933). *The Natural History of the Frilled shark <u>Chlamydoselachus anguineus</u>* (American museum of natural history. The Bashford Dean memorial volume, archaic fishes). Published by order of the trustees of the American Museum of Natural History, NY, USA.
Hannibal, J. T. (1997). *Remains of Arthropleura, a gigantic myriapod arthropod, from the Pennsylvanian of Ohio and Pennsylvania (Kirtlandia).* Cleveland Museum of Natural History, OH, USA.

Heuvelmans, B. (1958). *On the Track of Unknown Animals*. (3st ed). Kegan Paul, London, UK.
Heuvelmans, B. (1969). *In the Wake of the Sea Serpents*. (2nd ed). Hill and Wang, NY, USA.
Heywood, M., and Wells, S. (1995). *The Manual of Marine Invertebrates*. Voyageur Press, MN, USA.
Kirk, J. (1998). *In the Domain of the Lake Monsters: The Search for the Denizens of the Deep*. (1st ed). Key Porter Books, Toronto, Canada.
Kruuk, H. (2006). *Otters: Ecology, Behavior and Conservation*. Oxford University Press, NY, USA.
LeBlond, P. H., and Bousfield, E. L. (1995). *Cadborosaurus, Survivor from the Deep*. Heritage House Publishing, British Columbia, Canada.
Lee, H. (1883). *Sea Monsters Unmasked.* William Clowes and Sons Ltd, London, UK.
Lewis, O., Dowley, L., Marsh, S., Shuker, K. P. N., Downes, J., Moiser, C., Freeman, R., and Clark, C. (2006). *The Centre for Fortean Zoology Expedition Report 2006 – Gambia*. CFZ Press, Devon, UK.
Magnus, O. (1555). *History of the Northern Peoples*. Rome, Italy.
Mackal, R. P. (1976). *The Monsters of Loch Ness*. (Swallow Press, Chicago, USA.)
Mangiacopra, G. S., and Smith, D. G. (2007). *Does Champ Exist? Notes on the Historic Lake Monster Conference Held in Shelburne, Vermont, 29 August 1981*. Coachwhip Publications, PA, USA.
Meurger, M., and Gagnon, C. (1988). *Lake Monster Traditions: A Cross-Cultural Analysis*. Fortean Tomes, London, UK.
Moyle, P. B., and Cech Jnr, J. (2004) *Fishes: An Introduction to ichthyology*. Prentice-Hall Inc, NJ, USA.
Oudemans, A. C. (1892). *The Great Sea-Serpent*. (2007 reprint). Coachwhip Publications, PA, USA.
Peattie, N. (1996). *Hydra and Kraken, Or, the Lore and Lure of Lake-Monsters and Sea-Serpents*. Regent Press, CA, USA.
Riedman, M. (1991). *The Pinnipeds; Seals, Sea Lions and Walruses*. University of California Press, CA, USA
Rouse, G. W., and Pleijel, F. (2001). *Polychaetes*. Oxford University Press, New York, USA.
Shuker, K. P. N. (2003) *The Beasts That Hide From Man: Seeking the Worlds Last Undiscovered Animals.* Paraview Press, New York, USA.
Thewissen J.G.M (Ed., 1998) *The emergence of whales*; *evolutionary patterns in the origin of cetacean* (Plenum Press, New York and London)

Articles and chapters

Bajpai, S., and Thewissen, J. G. M. (2001). "A new, diminutive Eocene whale from Kachchh (Gujarat, India) and its implications for locomotor evolution of cetaceans." ***Current Science*. 79:10, 25, 1478– 1482.**
Berta, A. (1991). "New Enaliarctos* (Pinnipedimorpha) from the Miocene of Oregon and the role of "Enaliarctids" in Pinniped Phylogeny." ***Smithsonian Contributions to Paleobiology*. 69, 1-33.**

Bousfield, E. L., and LeBlond, P. H. (1995). "An account of *Cadborosaurus willsi*, new genus, new species, a large aquatic reptile from the Pacific coast of North America." *Amphipacifica: Journal of Systematic Biology.* 1:1, 1-25.

Buffetaut, E. (1982*)*. "Radiation évolutive, paléoécologie et biogéographie des Crocodiliens mésosuchienes." *Mémoires Societé Geologique de France.* **142**, 1–88.

Champagne, B. A. (2001). "A preliminary evaluation of a study of the morphology, behavior, autoecology and habitat of large unidentified marine animals, based on recorded field observations." *Crypto* Dracontology special 1. 93-112.

Champagne, B. A. (2007). "A classification system for large unidentified marine animals based on the examination of reported observations" in *Elementum Bestia* edited by **Heinselman C**, CRYPTO, Morrisville, NC, USA.144-172

Cope, E. D. (1874). "Review of the Vertebrata of the Cretaceous period found west of the Mississippi River." *U. S. Geological Survey of the Territories, Bulletin.* 1:2, 3-48.

Cornes, R. (2001). "The case of the surreal seal." *Crypto* Dracontology Special 1. 39-45.

Cornes, R. (2007). "The Seal Serpent: The Case for the Surreal Seal." In *The Centre for Fortean Zoology 2007 Yearbook*, edited by **Downes, J**. CFZ Press, Devon, UK. 83–199

Fraas, E. (1902). "Die Meer-Krocodilier (Thalattosuchia) des oberen Jura unter specieller Berücksichtigung von Dacosaurus und Geosaurus." *Paleontographica.* 49, 1-72.

Heuvelmans, B. (1988). "The sources and method of Cryptozoological research." *Cryptozoology,* 7, 1-21.

Kojo, Y. (1991). "Some ecological notes on reported large, unknown animals in lake Champlain." *Cryptozoology.* 10, 42-45.

Lazo, D. G., and Chichowolski, M. (2003). "First Plesiosaur Remains From the Lower Cretaceous of the Neuquén Basin, Argentina*."* *Journal of Paleontology.* 77:4, 784–789.

LeBlond, P. H. (1983). "An estimate of the dimensions of the Lake Champlain monster from the length of adjacent wind waves in the Mansi photograph." *Cryptozoology.* 1, 54-61.

LeBlond, P. H. (1993). "Sea Serpents of the Pacific Northwest." *Montana Mag* 43, 44-51.

Lee, M. S. Y. (1997). "The phylogeny of varanoid lizards and the affinities of snakes." *Philosophical Transactions of the Royal Society.* 352, 53-91.

Mangiacopra, G. S. (1992). "Theoretical Population Estimates of the Large Aquatic Animals in Selected Freshwater Lakes in North America." Thesis presented to Southern State University Connecticut.

Mangiacopra, G. S., Smith, A., and Avery, D. (2000). "Calculations of a size-density population of lake-monsters based upon a lake's physical limnology." *Crypto,* **3:2**, 5-11.

Motani, R., Minoura, N. and Ando, T. (1998). "Ichthyosaurian relationships illuminated by new primitive skeletons from Japan." *Nature.* 393, 255-257.

Muizon, C. de. (1981). "Les vertébrés fossiles de la formation Pisco (Pérou). Première partie: deux nouveaux Monachinae (Phocidae, Mammalia) du Pliocene de Sud-Sacaco." *Travaux de l'Institut Français d'Études Andines.* 22, 1-161.

Naish, D. (2001). "Sea serpents, seals and coelacanths: an attempt at a holistic approach to the identity of large aquatic cryptids." In *Fortean Studies Volume 7*, edited by Simmons, I. and Quin, M. John Brown Publishing, London, UK. 75-94

Naish, D. (2003). "CZ Conversations: Darren Naish on Plesiosaurs, Basilosaurs and Problems with reconstruction." *North American BioFortean Review.* **5, 3:12, 10-19.**

Paxton, C. G. M. (1998). "A cumulative species description curve for large, open water marine animals." *Journal of the Marine Biological Association of the United Kingdom.* 78:4, 1389.
Paxton, C. G. M., Knatterud, E., and Hedley, S.L. (2005). "Cetaceans, sex and sea serpents: an analysis of the Egede accounts of a "most dreadful monster" seen off the coast of Greenland in 1734." *Archives of Natural History.* 32, 1-9.
Radford, B. (2003). "The measure of a monster: investigating the Champ photo." *Skeptical Inquirer.* 19. Retrieved from: http://www.csicop.org/si/2003-07/monster.html on 24/02/07.
Rafinesque-Schmaltz, C. S. (1819). "Dissertation on Water-Snakes, Sea Snakes, and Sea Serpents." *The Philosophical Magazine and Journal.* 54, 361-367.
Thewissen, J. G. M., Madar, S.I., and Hussain, S. T. (1996). "*Ambulocetus natans*, an Eocene cetacean (Mammalia) from Pakistan". *Courier Forschungsinstitut Senckenberg.* 191, 1-86.
Thewissen, J. G. M., Williams, E. M., Roe, L. J., and Hussain, S. T. (2001). "Skeletons of terrestrial cetaceans and the relationship of whales to artiodactyls". *Nature,* 413, 277-281.
Woodley, M. A. (2008). "Towards a possible caudata identity for the Mongolian death worm: introducing the 'plausibility method' for identity theory formation amongst lesser known cryptids " In *The Centre for Fortean Zoology 2008 Yearbook*, edited by Downes, J., and Downes, **C. N. CFZ Press, Devon, UK. 39-47.**

Web sites

"**Cryptozoology: A Critical Approach**" **by Cameron McCormick.** At http://www.geocities.com/capedrevenger/index.html retrieved on 30/08/07.

"**Memories of Myspace past or the Cetacean Centipede Rides Again!**" by Cameron McCormick. At http://cameronmccormick.blogspot.com/2007/01/memories-of-myspace-past-or-cetacean.html retrieved on 10/10/2007.

"**Tetrapod Zoology**" **a blog by Darren Naish.** At http://scienceblogs.com/tetrapodzoology/ retrieved on 20/01/08.

"**The Centre for Fortean Zoology.**" At http://www.cfz.org.uk retrieved on 13/01/08.

"**The Great Sea Centipede**" by Lance Bradshaw. At http://www.angelfire.com/sc2/Trunko/centipede.html retrieved on 10/10/2007 retrieved on 10/10/2007.

"**The Plesiosaur Directory**" by Adam Stuart Smith. At http://www.plesiosauria.com/ retrieved on 31/08/07.

Image References and Copyright information

An introduction to the history and future of sea serpent classification

Fig. 1 The integrative method illustrated. My construction.
Fig. 2 The multifactor method illustrated. My construction.
Fig. 3 The plausibility method illustrated. My construction.

Ch. 1 – Heuvelmans' pinnipeds

Fig. 1 Plesiosaurus, by Adam Stuart Smith, 2002. Used under the Creative Commons Attribution ShareAlike 2.5 license.
Fig. 2 Giant squid 'posturing' from Lee, 1888. Public domain.
Fig. 3 *Megophias* from Ouedemans, 1892. Public domain.
Fig. 4 'Hydrarchos' from Koch, 1845. Public domain.
Fig. 5 Basilosaurus. Used under the GNU Free Documentation License.
Fig. 6 Pinniped cladogram. My construction.
Fig. 7 ***Enaliarctos*** from Wikipedia, Wikimedia Commons.
Fig. 8 ***Megolatoria longicollis*** from McCormick, 2006 © Modified from original – used with permission.
Fig. 9 ***Halshippus olai-magni*** from McCormick, 2006 © Modified from original – used with permission.
Fig. 10 Graph of Atlantic cod catch size vs. year. Data from Food and Agricultural Organization.
Fig. 11 Graph of longneck sightings vs. year. Data from Heuvelmans.

Fig. 12	Reconstructed swan necked seal from Naish, 2006 © Fair use.
Fig. 13	Graph of longneck sightings vs. coastal region. Data from Heuvelmans.
Fig. 14	Graph of *Megalotaria* and super-otter sightings vs. year. Data from Heuvelmans.
Fig. 15	Illustration of the 'Sea Orm' from Magnus, 1555. Public domain.
Fig. 16	Illustration of the *City of Baltimore* sighting from Ouedemans, 1892. **Fig.**
Fig. 11	Graph of longneck sightings vs. year. Data from Heuvelmans.
Fig. 12	Reconstructed swan necked seal from Naish, 2006 © Fair use.
Fig. 13	Graph of longneck sightings vs. coastal region. Data from Heuvelmans.
Fig. 14	Graph of *Megalotaria* and super-otter sightings vs. year. Data from Heuvelmans.
Fig. 15	Illustration of the 'Sea Orm' from Magnus, 1555. Public domain.
Fig. 16	Illustration of the *City of Baltimore* sighting from Ouedemans, 1892. Public domain.
Fig. 17	The Batchelor illustration from Heuvelmans, 1968. Public domain.
Fig. 18	The Mackintosh Bell sketches from Heuvelmans, 1968. Public domain.
Fig. 19	The Miller sketch from LeBlond and Bousfield, 1995 Unknown; fair use.
Fig. 20	Photograph of a 'Caddy' carcass taken from a whale, 1937 © British Columbia Archives and Record Service. Fair use.
Fig. 21	Photograph of 'Morgawr' taken by 'Mary F.', 1976 Unknown used with permission Tony Shiels
Fig. 22	Photograph of 'Champ' taken by Sandra Mansi, 1977 © Fair use.

Ch. 2 - The super-otter and the many-humped sea serpent: close cousins?

Fig. 1	The super-otter from McCormick, 2006 © Used with permission.
Fig. 2	The multiple humped sea serpent from McCormick, 2006 © Used with permission.
Fig. 3	*Ambulocetus*. From Wikipedia, used through Wiokimedia Commons
Fig. 4	*Kutchicetus*. From the *National Geographic Magazine, November 2001, The Evolution of Whales*, by Douglas H. Chadwick, Shawn Gould and Robert Clark - Re-illustrated for public access distribution by Sharon Mooney ©2006. Used under the Creative Commons Attribution ShareAlike 2.5 license.
Fig. 5	European otter, photographed by Catherine Trigg, 2007. Used under the Creative Commons Attribution 2.0 licence.

Ch. 3 – Others.
Pt. 1 - Marine 'saurians': genuine archaeocetes?

Fig. 1	Marine saurian from McCormick, 2006 © Used with permission.
Fig. 2	*Teleidosaurus*. Used under the GNU Free Documentation License Version 1.2.
Fig. 3	Mosasaurus. Public domain.

Fig. 4	*Rodhocetus*. From the *National Geographic Magazine*, November 2001, *The Evolution of Whales*, by Douglas H. Chadwick, Shawn Gould and Robert Clark - Re-illustrated for public access distribution by Sharon Mooney ©2006. Used under the Creative Commons Attribution ShareAlike 2.5 license.
Fig. 5	The *Sacramento* sea serpent, 1877. Public domain.
Fig. 6	'Gambo' carcass based on Burnham's sketches from CFZ.

Pt. 2 – Super-eels: a many faceted enigma.

Fig. 1	Frilled shark. Public domain.
Fig. 2	Oarfish. Public domain.
Fig. 3	The *Pauline* sea serpent. Public domain.

Pt. 3 - Giant invertebrates: The most plausible category.

Fig. 1	A salp colony string photographed by Lars Ploughmann, 2007. Used under the GNU Free Documentation License Version 1.2.
Fig. 2	Venus' girdle, Frull, 1998 © fair use.
Fig. 3	Lion's mane jellyfish. Public domain.
Fig. 4	Bootlace worm after Heuvelmans, 1968 © Fair use.

Michael Woodley has been an avid fan of Cryptozoology for as long as he can remember. He credits his early fascination with the discipline as being the catalyst for his life long love of Biology.

He has published articles in peer-reviewed journals on a range of subjects within the life-sciences, and is the author of *The Limits of Ecology: New Perspectives from a Theoretical Borderland* - a book of essays on Theoretical Ecology.

Michael holds a BSc degree from Columbia University, New York. He is currently studying for a PhD at the University of London, where he is investigating plant-bacteria interactions.

ACKNOWLEDGEMENTS

Firstly I would like to thank Jonathan Downes and the rest of the CFZ publishing tem, for being so tolerant of my whims and for doing such a fantastic job on both my 2008 Yearbook article and the front cover design for this book.

I would like to thank Chad Arment for his time in reviewing chapter one and for his helpful suggestions for improving the text. Also I would like to thank Cameron McCormick for allowing me the use of his fantastic images, and I would also like to thank Kenneth J. Sikes for his assistance in the editing and preparation of aspects of this manuscript. I would like to thank Dr Karl Shuker for his assistance with the final stages of editing. And finally I would like to thank my PhD supervisor, Paul Devlin, for always being there whenever I needed a sounding board for my more outlandish ideas, thanks!

THE CENTRE FOR FORTEAN ZOOLOGY

So, what is the Centre for Fortean Zoology?

We are a non profit-making organisation founded in 1992 with the aim of being a clearing house for information, and coordinating research into mystery animals around the world. We also study out of place animals, rare and aberrant animal behaviour, and Zooform Phenomena – little-understood "things" that appear to be animals, but which are in fact nothing of the sort, and not even alive (at least in the way we understand the term).

Why should I join the Centre for Fortean Zoology?

Not only are we the biggest organisation of our type in the world but - or so we like to think - we are the best. We are certainly the only truly global Cryptozoological research organisation, and we carry out our investigations using a strictly scientific set of guidelines. We are expanding all the time and looking to recruit new members to help us in our research into mysterious animals and strange creatures across the globe. Why should you join us? Because, if you are genuinely interested in trying to solve the last great mysteries of Mother Nature, there is nobody better than us with whom to do it.

What do I get if I join the Centre for Fortean Zoology?

For £12 a year, you get a four-issue subscription to our journal *Animals & Men*. Each issue contains 60 pages packed with news, articles, letters, research papers, field reports, and even a gossip column! The magazine is A5 in format with a full colour cover. You also have access to one of the world's largest collections of resource material dealing with cryptozoology and allied disciplines, and people from the CFZ membership regularly take part in fieldwork and expeditions around the world.

How is the Centre for Fortean Zoology organized?

The CFZ is managed by a three-man board of trustees, with a non-profit making trust registered with HM Government Stamp Office. The board of trustees is supported by a Permanent Directorate of full and part-time staff, and advised by a Consultancy Board of specialists - many of whom who are world-renowned experts in their particular field. We have regional representatives across the UK, the USA, and many other parts of the world, and are affiliated with other organisations whose aims and protocols mirror our own.

I am new to the subject, and although I am interested I have little practical knowledge. I don't want to feel out of my depth. What should I do?

Don't worry. We were *all* beginners once. You'll find that the people at the CFZ are friendly and approachable. We have a thriving forum on the website which is the hub of an ever-growing electronic community. You will soon find your feet. Many members of the CFZ Permanent Directorate started off as ordinary members, and now work full time chasing monsters around the world.

I have an idea for a project which isn't on your website. What do I do?

Write to us, e-mail us, or telephone us. The list of future projects on the website is not exhaustive. If you have a good idea for an investigation, please tell us. We may well be able to help.

How do I go on an expedition?

We are always looking for volunteers to join us. If you see a project that interests you, do not hesitate to get in touch with us. Under certain circumstances we can help provide funding for your trip. If you look on the future projects section of the website, you can see some of the projects that we have pencilled in for the next few years.

In 2003 and 2004 we sent three-man expeditions to Sumatra looking for Orang-Pendek - a semi-legendary bipedal ape. The same three went to Mongolia in 2005, and Guyana in 2007. All three members started off merely subscribers to the CFZ magazine.

Next time it could be you!

Project Kerinci, Sumatra - 2003
In search of the bipedal ape Orang Pendek

How is the Centre for Fortean Zoology funded?

We have no magic sources of income. All our funds come from donations, membership fees, works that we do for TV, radio or magazines, and sales of our publications and merchandise. We are always looking for corporate sponsorship, and other sources of revenue. If you have any ideas for fund-raising please let us know. However, unlike other cryptozoological organisations in the past, we do not live in an intellectual ivory tower. We are not afraid to get our hands dirty, and furthermore we are not one of those organisations where the membership have to raise money so that a privileged few can go on expensive foreign trips. Our research teams both in the UK and abroad, consist of a mixture of experienced and inexperienced personnel. We are truly a community, and work on the premise that the benefits of CFZ membership are open to all.

What do you do with the data you gather from your investigations and expeditions?

Reports of our investigations are published on our website as soon as they are available. Preliminary reports are posted within days of the project finishing.

Each year we publish a 200 page yearbook containing research papers and expedition reports too long to be printed in the journal. We freely circulate our information to anybody who asks for it.

Is the CFZ community purely an electronic one?

No. Each year since 2000 we have held our annual convention - the *Weird Weekend* - in Exeter. It is three days of lectures, workshops, and excursions. But most importantly it is a chance for members of the CFZ to meet each other, and to talk with the members of the permanent directorate in a relaxed and informal setting and preferably with a pint of beer in one hand. Starting in 2006 - the *Weird Weekend* has been held in the idyllic rural location of Woolsery in North Devon.

We are hoping to start up some regional groups in both the UK and the US which will have regular meetings, work together on research projects, and maybe have a mini convention of their own.

Since relocating to North Devon in 2005 we have become ever more closely involved with other community organisations, and we hope that this trend will continue. We also work closely with Police Forces across the UK as consultants for animal mutilation cases, and during 2006 we intend to forge closer links with the coastguard and other community services. We want to work closely with those who regularly travel into the Bristol Channel, so that if the recent trend of exotic animal visitors to our coastal waters continues, we can be out there as soon as possible.

We are building a Visitor's Centre in rural North Devon. This will not be open to the general public, but will provide a museum, a library and an educational resource for our members (currently over 400) across the globe. We are also planning a youth organisation which will involve children and young people in our activities.

Apart from having been the only Fortean Zoological organisation in the world to have consistently published material on all aspects of the subject for over a decade, we have achieved the following concrete results:

- Disproved the myth relating to the headless, so-called sea-serpent carcass of Durgan beach in Cornwall, 1975
- Disproved the story of the 1988 puma skull of Lustleigh Cleave
- Carried out the only in-depth research ever into mythos of the Cornish Owlman
- Made the first records of a tropical species of lamprey
- Made the first records of a luminous cave gnat larva in Thailand.
- Discovered a possible new species of British mammal - The Beech Marten.
- In 1994-6 carried out the first archival fortean zoological survey of Hong Kong.
- In the year 2000, CFZ theories where confirmed when an entirely new species of lizard was found resident in Britain.
- Identified the monster of Martin Mere in Lancashire as a giant wels catfish
- Expanded the known range of Armitage's skink in the Gambia by 80%
- Obtained photographic evidence of the remains of Europe's largest known pike
- Carried out the first ever in-depth study of the *ninki-nanka*
- Carried out the first attempt to breed Puerto Rican cave snails in captivity
- Were the first European explorers to visit the `lost valley` in Sumatra
- Made the first video recordings of a new species of scorpion in Guyana, 2007
- Brought the first evidence of red-faced Guyanese pygmies back to the UK in 2007
- Discovered a new colour morph of the rainbow boa

EXPEDITIONS & INVESTIGATIOINS TO DATE INCLUDE

- 1998 Puerto Rico, Florida, Mexico *(Chupacabras)*
- 1999 Nevada *(Bigfoot)*
- 2000 Thailand *(Giant snakes called nagas)*
- 2002 Martin Mere *(Giant catfish)*
- 2002 Cleveland *(Wallaby mutilation)*
- 2003 Bolam Lake *(BHM Reports)*
- 2003 Sumatra *(Orang Pendek)*
- 2003 Texas *(Bigfoot; giant snapping turtles)*
- 2004 Sumatra *(Orang Pendek; cigau, a sabre-toothed cat)*
- 2004 Illinois *(Black panthers; cicada swarm)*
- 2004 Texas *(Mystery blue dog)*
- 2004 Puerto Rico *(Chupacabras; carnivorous cave snails)*
- 2005 Belize *(Affiliate expedition for hairy dwarfs)*
- 2005 Mongolia *(Allghoi Khorkhoi aka Mongolian death worm)*
- 2006 Gambia *(Gambo - Gambian sea monster , Ninki Nanka and Armitage s skink*
- 2006 Llangorse Lake *(Giant pike, giant eels)*
- 2006 Windermere *(Giant eels)*
- 2007 Coniston Water *(Giant eels)*
- 2007 Guyana *(Giant anaconda, didi, water tiger)*

To apply for a <u>FREE</u> information pack about the organisation and details of how to join, plus information on current and future projects, expeditions and events.

Send a stamped and addressed envelope to:

**THE CENTRE FOR FORTEAN ZOOLOGY
MYRTLE COTTAGE, WOOLSERY,
BIDEFORD, NORTH DEVON
EX39 5QR.**

or alternatively visit our website at:
www.cfz.org.uk

Other books available from
CFZ PRESS

THE OWLMAN AND OTHERS - 30th Anniversary Edition
Jonathan Downes - ISBN 978-1-905723-02-7

£14.99

EASTER 1976 - Two young girls playing in the churchyard of Mawnan Old Church in southern Cornwall were frightened by what they described as a "nasty bird-man". A series of sightings that has continued to the present day. These grotesque and frightening episodes have fascinated researchers for three decades now, and one man has spent years collecting all the available evidence into a book. To mark the 30th anniversary of these sightings, Jonathan Downes has published a special edition of his book.

DRAGONS - More than a myth?
Richard Freeman - ISBN 0-9512872-9-X

£14.99

First scientific look at dragons since 1884. It looks at dragon legends worldwide, and examines modern sightings of dragon-like creatures, as well as some of the more esoteric theories surrounding dragonkind.

Dragons are discussed from a folkloric, historical and cryptozoological perspective, and Richard Freeman concludes that: "When your parents told you that dragons don't exist - they lied!"

MONSTER HUNTER
Jonathan Downes - ISBN 0-9512872-7-3

£14.99

Jonathan Downes' long-awaited autobiography, *Monster Hunter*...

Written with refreshing candour, it is the extraordinary story of an extraordinary life, in which the author crosses paths with wizards, rock stars, terrorists, and a bewildering array of mythical and not so mythical monsters, and still just about manages to emerge with his sanity intact.......

MONSTER OF THE MERE
Jonathan Downes - ISBN 0-9512872-2-2

£12.50

It all starts on Valentine's Day 2002 when a Lancashire newspaper announces that "Something" has been attacking swans at a nature reserve in Lancashire. Eyewitnesses have reported that a giant unknown creature has been dragging fully grown swans beneath the water at Martin Mere. An intrepid team from the Exeter based Centre for Fortean Zoology, led by the author, make two trips – each of a week – to the lake and its surrounding marshlands. During their investigations they uncover a thrilling and complex web of historical fact and fancy, quasi Fortean occurrences, strange animals and even human sacrifice.

**CFZ PRESS, MYRTLE COTTAGE,
WOOLFARDISWORTHY BIDEFORD,
NORTH DEVON, EX39 5QR
www.cfz.org.uk**

Other books available from
CFZ PRESS

ONLY FOOLS AND GOATSUCKERS
Jonathan Downes - ISBN 0-9512872-3-0

£12.50

In January and February 1998 Jonathan Downes and Graham Inglis of the Centre for Fortean Zoology spent three and a half weeks in Puerto Rico, Mexico and Florida, accompanied by a film crew from UK Channel 4 TV. Their aim was to make a documentary about the terrifying chupacabra - a vampiric creature that exists somewhere in the grey area between folklore and reality. This remarkable book tells the gripping, sometimes scary, and often hilariously funny story of how the boys from the CFZ did their best to subvert the medium of contemporary TV documentary making and actually do their job.

WHILE THE CAT'S AWAY
Chris Moiser - ISBN: 0-9512872-1-4

£7.99

Over the past thirty years or so there have been numerous sightings of large exotic cats, including black leopards, pumas and lynx, in the South West of England. Former Rhodesian soldier Sam McCall moved to North Devon and became a farmer and pub owner when Rhodesia became Zimbabwe in 1980. Over the years despite many of his pub regulars having seen the "Beast of Exmoor" Sam wasn't at all sure that it existed. Then a series of happenings made him change his mind. Chris Moiser—a zoologist—is well known for his research into the mystery cats of the westcountry. This is his first novel.

CFZ EXPEDITION REPORT 2006 - GAMBIA
ISBN 1905723032

£12.50

In July 2006, The J.T.Downes memorial Gambia Expedition - a six-person team - Chris Moiser, Richard Freeman, Chris Clarke, Oll Lewis, Lisa Dowley and Suzi Marsh went to the Gambia, West Africa. They went in search of a dragon-like creature, known to the natives as `Ninki Nanka`, which has terrorized the tiny African state for generations, and has reportedly killed people as recently as the 1990s. They also went to dig up part of a beach where an amateur naturalist claims to have buried the carcass of a mysterious fifteen foot sea monster named 'Gambo', and they sought to find the Armitage's Skink (*Chalcides armitagei*) - a tiny lizard first described in 1922 and only rediscovered in 1989. Here, for the first time, is their story.... With an forward by Dr. Karl Shuker and introduction by Jonathan Downes.

BIG CATS IN BRITAIN YEARBOOK 2006
Edited by Mark Fraser - ISBN 978-1905723-01-0

£10.00

Big cats are said to roam the British Isles and Ireland even now as you are sitting and reading this. People from all walks of life encounter these mysterious felines on a daily basis in every nook and cranny of these two countries. Most are jet-black, some are white, some are brown, in fact big cats of every description and colour are seen by some unsuspecting person while on his or her daily business. 'Big Cats in Britain' are the largest and most active group in the British Isles and Ireland This is their first book. It contains a run-down of every known big cat sighting in the UK during 2005, together with essays by various luminaries of the British big cat research community which place the phenomenon into scientific, cultural, and historical perspective.

**CFZ PRESS, MYRTLE COTTAGE,
WOOLSERY, BIDEFORD,
NORTH DEVON, EX39 5QR
www.cfz.org.uk**

Other books available from
CFZ PRESS

THE SMALLER MYSTERY CARNIVORES OF THE WESTCOUNTRY
Jonathan Downes - ISBN 978-1-905723-05-8

£7.99

Although much has been written in recent years about the mystery big cats which have been reported stalking Westcountry moorlands, little has been written on the subject of the smaller British mystery carnivores. This unique book redresses the balance and examines the current status in the Westcountry of three species thought to be extinct: the Wildcat, the Pine Marten and the Polecat, finding that the truth is far more exciting than the currently held scientific dogma. This book also uncovers evidence suggesting that even more exotic species of small mammal may lurk hitherto unsuspected in the countryside of Devon, Cornwall, Somerset and Dorset.

THE BLACKDOWN MYSTERY
Jonathan Downes - ISBN 978-1-905723-00-3

£7.99

Intrepid members of the CFZ are up to the challenge, and manage to entangle themselves thoroughly in the bizarre trappings of this case. This is the soft underbelly of ufology, rife with unsavoury characters, plenty of drugs and booze." That sums it up quite well, we think. A new edition of the classic 1999 book by legendary fortean author Jonathan Downes. In this remarkable book, Jon weaves a complex tale of conspiracy, anti-conspiracy, quasi-conspiracy and downright lies surrounding an air-crash and alleged UFO incident in Somerset during 1996. However the story is much stranger than that. This excellent and amusing book lifts the lid off much of contemporary forteana and explains far more than it initially promises.

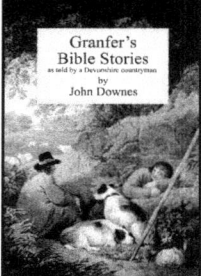

GRANFER'S BIBLE STORIES
John Downes - ISBN 0-9512872-8-1

£7.99

Bible stories in the Devonshire vernacular, each story being told by an old Devon Grandfather - 'Granfer'. These stories are now collected together in a remarkable book presenting selected parts of the Bible as one more-or-less continuous tale in short 'bite sized' stories intended for dipping into or even for bed-time reading. `Granfer` treats the biblical characters as if they were simple country folk living in the next village. Many of the stories are treated with a degree of bucolic humour and kindly irreverence, which not only gives the reader an opportunity to re-evaluate familiar tales in a new light, but do so in both an entertaining and a spiritually uplifting manner.

FRAGRANT HARBOURS DISTANT RIVERS
John Downes - ISBN 0-9512872-5-7

£12.50

Many excellent books have been written about Africa during the second half of the 19th Century, but this one is unique in that it presents the stories of a dozen different people, whose interlinked lives and achievements have as many nuances as any contemporary soap opera. It explains how the events in China and Hong Kong which surrounded the Opium Wars, intimately effected the events in Africa which take up the majority of this book. The author served in the Colonial Service in Nigeria and Hong Kong, during which he found himself following in the footsteps of one of the main characters in this book; Frederick Lugard – the architect of modern Nigeria.

CFZ PRESS, MYRTLE COTTAGE, WOOLFARDISWORTHY BIDEFORD, NORTH DEVON, EX39 5QR
w w w . c f z . o r g . u k

Other books available from
CFZ PRESS

ANIMALS & MEN - Issues 1 - 5 - In the Beginning
Edited by Jonathan Downes - ISBN 0-9512872-6-5

£12.50

At the beginning of the 21st Century monsters still roam the remote, and sometimes not so remote, corners of our planet. It is our job to search for them. The Centre for Fortean Zoology [CFZ] is the only professional, scientific and full-time organisation in the world dedicated to cryptozoology - the study of unknown animals. Since 1992 the CFZ has carried out an unparalleled programme of research and investigation all over the world. We have carried out expeditions to Sumatra (2003 and 2004), Mongolia (2005), Puerto Rico (1998 and 2004), Mexico (1998), Thailand (2000), Florida (1998), Nevada (1999 and 2003), Texas (2003 and 2004), and Illinois (2004). An introductory essay by Jonathan Downes, notes putting each issue into a historical perspective, and a history of the CFZ.

ANIMALS & MEN - Issues 6 - 10 - The Number of the Beast
Edited by Jonathan Downes - ISBN 978-1-905723-06-5

£12.50

At the beginning of the 21st Century monsters still roam the remote, and sometimes not so remote, corners of our planet. It is our job to search for them. The Centre for Fortean Zoology [CFZ] is the only professional, scientific and full-time organisation in the world dedicated to cryptozoology - the study of unknown animals. Since 1992 the CFZ has carried out an unparalleled programme of research and investigation all over the world. We have carried out expeditions to Sumatra (2003 and 2004), Mongolia (2005), Puerto Rico (1998 and 2004), Mexico (1998), Thailand (2000), Florida (1998), Nevada (1999 and 2003), Texas (2003 and 2004), and Illinois (2004). Preface by Mark North and an introductory essay by Jonathan Downes, notes putting each issue into a historical perspective, and a history of the CFZ.

BIG BIRD! Modern Sightings of Flying Monsters

Ken Gerhard - ISBN 978-1-905723-08-9

£7.99

From all over the dusty U.S./Mexican border come hair-raising stories of modern day encounters with winged monsters of immense size and terrifying appearance. Further field sightings of similar creatures are recorded from all around the globe. What lies behind these weird tales? Ken Gerhard is a native Texan, he lives in the homeland of the monster some call 'Big Bird'. Ken's scholarly work is the first of its kind. On the track of the monster, Ken uncovers cases of animal mutilations, attacks on humans and mounting evidence of a stunning zoological discovery ignored by mainstream science. Keep watching the skies!

STRENGTH THROUGH KOI
They saved Hitler's Koi and other stories

Jonathan Downes - ISBN 978-1-905723-04-1

£7.99

Strength through Koi is a book of short stories - some of them true, some of them less so - by noted cryptozoologist and raconteur Jonathan Downes. The stories are all about koi carp, and their interaction with bigfoot, UFOs, and Nazis. Even the late George Harrison makes an appearance. Very funny in parts, this book is highly recommended for anyone with even a passing interest in aquaculture, but should be taken definitely *cum grano salis*.

CFZ PRESS, MYRTLE COTTAGE,
WOOLSERY, BIDEFORD,
NORTH DEVON, EX39 5QR

Other books available from
CFZ PRESS

BIG CATS IN BRITAIN YEARBOOK 2007
Edited by Mark Fraser - ISBN 978-1-905723-09-6

£12.50

People from all walks of life encounter mysterious felids on a daily basis, in every nook and cranny of the UK. Most are jet-black, some are white, some are brown; big cats of every description and colour are seen by some unsuspecting person while on his or her daily business. 'Big Cats in Britain' are the largest and most active research group in the British Isles and Ireland. This book contains a run-down of every known big cat sighting in the UK during 2006, together with essays by various luminaries of the British big cat research community.

CAT FLAPS! Northern Mystery Cats
Andy Roberts - ISBN 978-1-905723-11-9

£6.99

Of all Britain's mystery beasts, the alien big cats are the most renowned. In recent years the notoriety of these uncatchable, out-of-place predators have eclipsed even the Loch Ness Monster. They slink from the shadows to terrorise a community, and then, as often as not, vanish like ghosts. But now film, photographs, livestock kills, and paw prints show that we can no longer deny the existence of these once-legendary beasts. Here then is a case-study, a true lost classic of Fortean research by one of the country's most respected researchers.

CENTRE FOR FORTEAN ZOOLOGY 2007 YEARBOOK
Edited by Jonathan Downes and Richard Freeman
ISBN 978-1-905723-14-0

£12.50

The Centre For Fortean Zoology Yearbook is a collection of papers and essays too long and detailed for publication in the CFZ Journal *Animals & Men*. With contributions from both well-known researchers, and relative newcomers to the field, the Yearbook provides a forum where new theories can be expounded, and work on little-known cryptids discussed.

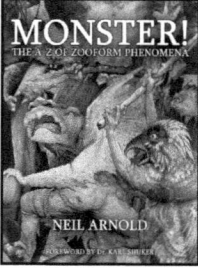

MONSTER! THE A-Z OF ZOOFORM PHENOMENA
Neil Arnold - ISBN 978-1-905723-10-2

£14.99

Zooform Phenomena are the most elusive, and least understood, mystery `animals`. Indeed, they are not animals at all, and are not even animate in the accepted terms of the word. Author and researcher Neil Arnold is to be commended for a groundbreaking piece of work, and has provided the world's first alphabetical listing of zooforms from around the world.

**CFZ PRESS, MYRTLE COTTAGE,
WOOLFARDISWORTHY BIDEFORD,
NORTH DEVON, EX39 5QR
w w w . c f z . o r g . u k**

Other books available from
CFZ PRESS

BIG CATS LOOSE IN BRITAIN
Marcus Matthews - ISBN 978-1-905723-12-6

£14.99

Big Cats: Loose in Britain, looks at the body of anecdotal evidence for such creatures: sightings, livestock kills, paw-prints and photographs, and seeks to determine underlying commonalities and threads of evidence. These two strands are repeatedly woven together into a highly readable, yet scientifically compelling, overview of the big cat phenomenon in Britain.

DARK DORSET
TALES OF MYSTERY, WONDER AND TERROR
Robert. J. Newland and Mark. J. North
ISBN 978-1-905723-15-6

£12.50

This extensively illustrated compendium has over 400 tales and references, making this book by far one of the best in its field. Dark Dorset has been thoroughly researched, and includes many new entries and up to date information never before published. The title of the book speaks for itself, and is indeed not for the faint hearted or those easily shocked.

MAN-MONKEY - IN SEARCH OF THE BRITISH BIGFOOT
Nick Redfern - ISBN 978-1-905723-16-4

£9.99

In her 1883 book, *Shropshire Folklore*, Charlotte S. Burne wrote: *'Just before he reached the canal bridge, a strange black creature with great white eyes sprang out of the plantation by the roadside and alighted on his horse's back'*. The creature duly became known as the `Man-Monkey`.

Between 1986 and early 2001, Nick Redfern delved deeply into the mystery of the strange creature of that dark stretch of canal. Now, published for the very first time, are Nick's original interview notes, his files and discoveries; as well as his theories pertaining to what lies at the heart of this diabolical legend.

EXTRAORDINARY ANIMALS REVISITED
Dr Karl Shuker - ISBN 978-1905723171

£14.99

This delightful book is the long-awaited, greatly-expanded new edition of one of Dr Karl Shuker's much-loved early volumes, *Extraordinary Animals Worldwide*. It is a fascinating celebration of what used to be called romantic natural history, examining a dazzling diversity of animal anomalies, creatures of cryptozoology, and all manner of other thought-provoking zoological revelations and continuing controversies down through the ages of wildlife discovery.

CFZ PRESS, MYRTLE COTTAGE,
WOOLFARDISWORTHY BIDEFORD,
NORTH DEVON, EX39 5QR
w w w . c f z . o r g . u k

Other books available from
CFZ PRESS

DARK DORSET CALENDAR CUSTOMS
Robert J Newland - ISBN 978-1-905723-18-8

£12.50

Much of the intrinsic charm of Dorset folklore is owed to the importance of folk customs. Today only a small amount of these curious and occasionally eccentric customs have survived, while those that still continue have, for many of us, lost their original significance. Why do we eat pancakes on Shrove Tuesday? Why do children dance around the maypole on May Day? Why do we carve pumpkin lanterns at Hallowe'en? All the answers are here! Robert has made an in-depth study of the Dorset country calendar identifying the major feast-days, holidays and celebrations when traditionally such folk customs are practiced.

CENTRE FOR FORTEAN ZOOLOGY 2004 YEARBOOK
Edited by Jonathan Downes and Richard Freeman
ISBN 978-1-905723-14-0

£12.50

The Centre For Fortean Zoology Yearbook is a collection of papers and essays too long and detailed for publication in the CFZ Journal *Animals & Men*. With contributions from both well-known researchers, and relative newcomers to the field, the Yearbook provides a forum where new theories can be expounded, and work on little-known cryptids discussed.

CENTRE FOR FORTEAN ZOOLOGY 2008 YEARBOOK
Edited by Jonathan Downes and Corinna Downes
ISBN 978 -1-905723-19-5

£12.50

The Centre For Fortean Zoology Yearbook is a collection of papers and essays too long and detailed for publication in the CFZ Journal *Animals & Men*. With contributions from both well-known researchers, and relative newcomers to the field, the Yearbook provides a forum where new theories can be expounded, and work on little-known cryptids discussed.

ETHNA'S JOURNAL
Corinna Newton Downes
ISBN 978 -1-905723-21-8

£9.99

Ethna's Journal tells the story of a few months in an alternate Dark Ages, seen through the eyes of Ethna, daughter of Lord Edric. She is an unsophisticated girl from the fortress town of Cragnuth, somewhere in the north of England, who reluctantly gets embroiled in a web of treachery, sorcery and bloody war...

**CFZ PRESS, MYRTLE COTTAGE,
WOOLFARDISWORTHY BIDEFORD,
NORTH DEVON, EX39 5QR
w w w . c f z . o r g . u k**

Other books available from
CFZ PRESS

ANIMALS & MEN - Issues 11 - 15 - The Call of the Wild
Jonathan Downes (Ed) - ISBN 978-1-905723-07-2

£12.50

Since 1994 we have been publishing the world's only dedicated cryptozoology magazine, *Animals & Men*. This volume contains fascimile reprints of issues 11 to 15 and includes articles covering out of place walruses, feathered dinosaurs, possible North American ground sloth survival, the theory of initial bipedalism, mystery whales, mitten crabs in Britain, Barbary lions, out of place animals in Germany, mystery pangolins, the barking beast of Bath, Yorkshire ABCs, Molly the singing oyster, singing mice, the dragons of Yorkshire, singing mice, the bigfoot murders, waspman, British beavers, the migo, Nessie, the weird warbling whatsit of the westcountry, the quagga project and much more...

IN THE WAKE OF BERNARD HEUVELMANS
Michael A Woodley - ISBN 978-1-905723-20-1

£9.99

Everyone is familiar with the nautical maps from the middle ages that were liberally festooned with images of exotic and monstrous animals, but the truth of the matter is that the *idea* of the sea monster is probably as old as humankind itself.

For two hundred years, scientists have been producing speculative classifications of sea serpents, attempting to place them within a zoological framework. This book looks at these successive classification models, and using a new formula produces a sea serpent classification for the 21st Century.

**CFZ PRESS, MYRTLE COTTAGE,
WOOLFARDISWORTHY BIDEFORD,
NORTH DEVON, EX39 5QR
w w w . c f z . o r g . u k**

www.ingramcontent.com/pod-product-compliance
Ingram Content Group UK Ltd.
Pitfield, Milton Keynes, MK11 3LW, UK
UKHW021320180426
11947UKWH00015B/1341